The IF Trilogy
A Unified Theory of God, Mind, and Matter

By
L. R. Caldwell

Volume I — Theology
Volume II — Philosophy
Volume III — Science

Reason and Reality Publishing

Copyright © 2025 by L. R. Caldwell

All rights reserved. No part of this book may be reproduced, stored in a retrieval system, or transmitted in any form or by any means—electronic, mechanical, photocopying, recording, or otherwise—without prior written permission of the author, except in the case of brief quotations embodied in critical articles and reviews.

First Edition
ISBN: 979-8-9992710-1-3
https://orcid.org/0009-0005-6487-9274

Published by Reason and Reality Publishing

Florida, United States
This work is a product of independent research and philosophical inquiry.

While every effort has been made to ensure accuracy, the author makes no claims beyond the stated scope of the work.

Printed in the United States of America

Dedication

To my daughter, Chrysta C (Caldwell) Feely, as with my first book,
Consciousness: Beyond the Planck Boundary.

For *The IF Trilogy*:

Volume 1 — To my daughter Nicole M Caldwell

Volume 2 — To my son Yizhu JS Caldwell

Volume 3 — To my daughter Rachel L. Caldwell

They say traits are inherited, a phrase usually meant to describe qualities passed down through the generations.
Yet I truly believe traits can also be handed *up* through the generations.
Each of these children are unique in their own special way, and each has given me a distinct and personal perception of my work, my life, and the world.
At times, I think more has been inherited upward from them than I may have ever handed down.

A special thank you must also go out to two very special women-
My sister - Daphne L (Caldwell)Bagdonas
And
My lovely wife, Amy Caldwell

Contents

Chapter-1 The Origin of All Things: Consciousness Before Creation 1
  Scriptural Anchor 1
  CSFT Integration 1
  Comparison with Science 2
  Theological Alignment 3
  Chapter 2 5

Beyond the Planck Boundary: 5
  1. Introduction: The Problem of the Edge 5
  3. CSFT's Proposal: A Consciousness Field Preceding Physics 5
  4. Scriptural Alignment: Sacred Texts at the Edge of Physics 6
  Chapter 3 11

Language as Threshold and Filter 11
  The Power and the Problem of Language 11
  CSFT and the Need for Precise Language 12
  Scripture and the Risk of Literalism 14
  Conclusion: Humility Before the Word 14
  Chapter 4 17
  CSFT Response to Logical Necessity 17
  Respectful Theological Reframing 18
  Scriptural Alignment with CSFT 18
  Rejection of Structure as Rejection of God ... 20

Chapter 5 23
The Living Word and the Field of Meaning 23
- The Meaning of Logos 23
- The Power of Sacred Language 24
- A Unified Voice 25
- Chapter 6 29
- Sacred Narrative and the Field of Resonance (CSFT Perspective) 29
  - 3. Field Logic and Prophetic Initiation 30
  - Linguistic Resonance and Divine Language . 31
  - 8. Parables and Multilayered Meaning 32
  - Conclusion: The Divine Framework 33
- Chapter 7 35
  - 1. Resurrection as Field Logic 35
  - 3. Paul's Logical Argument for a Renewed Body 36
  - 6. Divine Coherence and the Transfiguration 38
  - 9. Resurrection in the Hebrew Bible 39
  - 11. The Role of Faith as Resonance Activation 40
- Chapter 8 43
- The Spirit and the Field 43
  - 1. Spirit as Breath: The Origin of Life 43
  - 4. Inspiration and Prophecy: Structured Reception 46

6. The Comforter: Logic in Times of Chaos..47
7. Spirit and Truth: Coherence as Holiness In 48
10. Breath and Spirit in Death and Life 50
Chapter 9 55
Covenant and Coherence – Walking with God through Structured Alignment..............................55
1. Covenant as Structural Realignment.........55
3. Moses and the Covenant of Law................56
6. Christ and the Covenant of Inner Coherence 57
9. Grace as Coherence Recovery 58
12. Conclusion: Covenant as the Path to Coherence..59
Chapter 10 61
Sin and Structural Collapse – Understanding Separation through Dissonance 61
1. What Is Sin in CSFT?.................................61
2. Coherence: The Opposite of Sin.................62
7. Intentional vs. Unintentional Sin..............64
8. Repentance and Structural Re-entry.........64
11. Forgiveness as Field Rebalance 65
Chapter 11 69
Redemption and Resonance – The Restoration of Structure through Divine Logic 69
Resonance and Dissonance in CSFT.............70

Renewal and Reidentification 71
Chapter 12 75
Free Will and Resonant Choice 75
The Modern Objection Against Free Will 75
The CSFT Response: A Third Way 76
Scriptural Alignment 77
Chapter 13 81
Covenant and the Logic of Collective Resonance 81
The Covenant as Structure, Not Superstition 81
The Invitation of Covenant Today 84
Chapter 14 87
The Living Logic of Faith 87
Faith as Alignment, Not Blindness 88
An Invitation to Continue the Journey 90

Author's Thoughts on Book 1 91

Chapter 1 93
1.1 The Bridge from Theology to Philosophy 93
1.2 Covenant as a Logical Structure 94
1.3 Coherence as the Philosophical Continuation of Faith 94
1.4 Setting the Stage for Book 2 95
Chapter 2 97
Section 1 – From Foundation to Flow 97
Section 2 – The Architecture of Structured Resonance 98

Section 3 – The Principle of Sufficient Reason in the Field 100
Section 4 – Toward the Next Horizon 101

Bibliography...102

Chapter 3 103
Genius & the Monad.....................................103
Chapter 4 107
Section 1 – From Metaphysical Premise to Operational Model 107
Section 2 – Differentiated Resonance...........107
Section 3 – The Excitation Principle (Expanded)..108
Section 4 – Boundaries Without Walls 110
Section 5 – Transition to Inquiry 111
Chapter 5 115
Inquiry at the Resonance Frontier Section 1 – Philosophy Meets Method 115
Section 3 – Designing Inquiry for a Consciousness-First Framework 118
Section 4 – Interpreting Data Through the Lens of Resonance 120
Section 5 – The Path Forward 121
Chapter 6 127

Leibniz, Kant, Whitehead, and CSFT with Critiques 127
Leibniz and CSFT 127

Kant and CSFT 128
Whitehead and CSFT 129
  Philosophical critiques of CSFT and responses
.... 130
    Leibnizian critique. 130
    Kantian critique. 130
    Whiteheadian critique. 130

  Bibliography 131

    Chapter 1 133
    Introduction to the Scientific Lens of the Trilogy 133
    Chapter 2 137
    Consciousness Structured Field Theory — Core Principles & Scientific Alignment 137
    Chapter 3 145

CSFT: Theoretical Integrity, Scientific Alignment, and Metaphysical Necessity 145

Section 1: Introduction and Methodological Foundations 146

Section 2: Logical Structure and Ontological Necessity 149

  2.3 Internal Coherence and Non-Circularity
.... 150
  2.4 Philosophical Parallels and Validation .. 150

Section 3: Scientific Parallels and Field Boundary Alignment 151
    3.1 Planck-Scale Boundary and Metaphysical Necessity 151
    3.2 Alignment with Quantum Mind Critiques 152
    3.3 Structural Resonance and Information Realism 152
Section 4: Implications, Predictive Power, and Academic Positioning 153
    4.2 Predictive and Theoretical Consequences 153
    4.3 Academic Positioning and Future Engagement 154
    Concluding References 155
    Chapter 4 157
CSFT: Today Philosophy, Tomorrow Scientific Theory? 157
    1. Introduction 157
    2. Historical Precedents of Ridiculed Philosophy Becoming Foundational Science 158
    2.2 Heliocentrism 159
    2.3 Germ Theory of Disease 159
    3. Linking CSFT to Historical Patterns 160
    4. Addressing Likely Criticisms 160

Chapter 5 - pt 1 165

Why Neuroscience May Never Solve Consciousness: A Field-Based Resolution to the Hard Problem 165

INTRO 165
1. The Ontological Failure of Neuroscience 165
2. The Hard Problem and the Limits of Correlation 165
3. Introducing the Consciousness-Structured Field Theory (CSFT) 166
4. Materialism and the Explanatory Gap 166
5. CSFT as a Structuring Alternative to Reductionism 167
6. Salt, Structure, and the Unity of Experience
.... 167
7. Addressing Critiques of CSFT 168
8. A New Paradigm for Consciousness Science
.... 168
Simplified Summary 168
References 169
Chapter 5 - pt 2 171

Original Insight: Neuroscience vs. CSFT Interpretation 171

1) Insight's distinctive neural dynamics 171

2) Incubation (Sio & Ormerod, 2009) and the default-mode (Beaty et al., 2015)/control partnership 172

3) "Restructuring" via coherence (Fries, 2005) = access routing, not authorship 172

4) The Aha! feels good—reward tagging vs. origin of content 172

5) Consciousness capacity, integration, and complexity 173

6) Addressing neuroscientific critiques (bridge requirements) 173

Bibliography 173

Chapter 6 175

Resonance as a Physical Principle in CSFT 175

2. A Brief History of Resonance in Science . 175

3. Why Resonance Matters in Science 176

4. Resonance Across Physics, Biology, and Neuroscience 177

5. The CSFT Connection: Resonance as a Brain–Field Interface 177

5.1. QFT Resonance: Poles, Widths, and Optimal Coupling 178

5.2. Mode/Phase Matching and Density-of-States Heuristics 178

7. Conclusion 180

Chapter 7 185

Thermodynamics, Information, and the CSFT Interface 185
   1. Introduction — From Energy to Information 185
   2. The Thermodynamic Foundation 185
   3. Local Excitation Hypothesis (CSFT) 186
   4. CSFT Interpretation 186
   5. Testable Predictions 187
   6. Positioning Against Critiques 187
   7. Conclusion — The Physics Anchor for CSFT 188
   Bibliography 188
   Chapter 8 191
Flatness, Relativity, and the Planck Boundary 191
   Summary 191
   1. Introduction: The Puzzle of a Flat Universe 192
   2. Inflation as the Conventional Solution 192
   3. Observational Limits and the Planck Boundary 193
   4. Einstein, Relativity, and the Observer 194
   5. Reframing the Flatness Problem 194
   Chapter 9 197
The Limits of Scientific Method 197

1. Introduction 197
2. The Strengths of the Scientific Method ... 197
3. Reductionism and Its Limits 197
4. Measurement and the Boundaries of Science 198
5. Materialism as an Assumption 198
6. Conclusion 199
Chapter 10 201

Consciousness and Objectivity 201

1. Introduction 201
2. The Ideal of Objectivity 201
3. Einstein and Relativity 201
4. Quantum Measurement and the Observer 202
5. CSFT and Structured Observation 202
6. Rethinking Objectivity 203
Chapter 11 205

Consciousness and Truth 205

1. Introduction 205
2. Classical Notions of Truth 206
3. Truth in Scientific Realism (Psillos, 1999) 206
4. Limits of Knowledge 207
5. CSFT and Truth as Structured Resonance 208
6. Implications for Science and Philosophy . 208

7. Conclusion 210
Chapter 12 215

Synthesis and Outlook 215

1. Purpose and Scope of the Conclusion 215
2. What Has Been Shown 215
3. Interfaces with Established Science 216
4. Limits, Risk, and Falsifiability 217
5. Programmatic Probes (Non-Medical, Domain-General) 218
6. Implications for Measurement and Explanation 219
7. Conclusion 219

CSFT Master Glossary 223

Beyond Neurosufficiency 223
Boundary Transcendence 223
Brute Force 223
Cognitive Resonance 223
Coherence Field 223
Coherent Alignment 223
Cold Resonance 224
Conscious Coherence 224
Conscious Excitation 224
Conscious System 224
Consciousness-Structured Field (CSF) 224
Deep Coherence 225
Emergent Resonance 225

Empirical 226
Entropy Gradient (CSFT Interpretation) ... 226
Excitation (in CSFT) 226
Excitation Threshold 226
Field Anchoring 226
Field Logic 226
Field of Consciousness 227
Field Participation 227
Foundational Symmetry 227
Gödelian Grounding 227
Harmonic Node 227
Logical Resonance 227
Logical Rules Encoded in Neurons 228
Mathematical Coherence (CSFT View) 228
Measurement Barrier 228
Metaphysical Substrate 228
Monad 228
Monadic Layering 228
Monadic Perspective 229
Monads (in CSFT) 229
Non-Emergent Logic 229
Non-Local Resonance 229
Ontological Asymmetry 229
Ontological Grounding 229
Planck Boundary 230
Planck Mirror Theory 230
Pre-Logical Order 230

Qualia 230
The subjective, first-person experiences that emerge from resonance with the consciousness field. In CSFT, qualia are the signatures of structured excitation within a conscious system 230
- **Quantum-Patterned Cosmos (QPC)** 230
- **Resonance Collapse** 231
- **Resonance Principles** 231
- **Resonance Stabilization** 231
- **Resonant Ground** 231
- **Resonant Interference** 231
- **Resonant Node** 231
- **SAR (System Aligned in Resonance)** 232
- **Structured Coherence** 232
- **Structured Differentiation** 232
- **Structured Potential** 232
- **Structured Resonance** 232
- **Sub-Planckian Layer** 232
- **Symmetries** 233
- **Transcendent Coherence** 233
- **Transcendent Logic** 233
- **Unity of Coherence** 233

Author's Words 235
Epilogue 237
- **A Unified Vision** 237

If - Volume 1 - Theology

Chapter-1
The Origin of All Things: Consciousness Before Creation

Opening Statement
The Origin of All Things: Consciousness Before Creation

Before time ticked, before space expanded, before light ever broke the silence, there was consciousness. Not the mind of man, but the primordial structure of awareness itself—eternal, logical, unshaped by matter. This is the foundational claim of the Consciousness-Structured Field Theory (CSFT): that consciousness is not a byproduct of creation, but its cause. In this chapter, we explore how this view not only supports sacred scripture but completes it.

Scriptural Anchor
"In the beginning God created the heaven and the earth... And the Spirit of God moved upon the face of the waters." (Genesis 1:1–2, KJV)
"In the beginning was the Word, and the Word was with God, and the Word was God." (John 1:1, KJV)

These verses do not describe the beginning of God—they describe the beginning of form. The "Spirit", or "Word", is present before structure, before light, before time. CSFT affirms this: the field of consciousness precedes the physical universe.

CSFT Integration

CSFT posits that all structure—matter, energy, time, and even the laws of physics—emerge from a deeper, logic-bound field of consciousness. This field is not passive.

It contains patterned regions capable of excitation, and it is from these structured nodes that existence takes form. "The Spirit of God, the Word, the Logos"—each point to this pre-structured field that underlies and informs reality.

This field is the basis of structured resonance—an organizing principle that allows existence to emerge with order and coherence, not randomness.

Logical Framing

The Principle of Sufficient Reason demands that for every effect, there must be a cause. But brute matter cannot explain its own origin. Something non-material, eternal, and structured must exist as the precondition for anything else to exist.

That something is consciousness. CSFT identifies this as a field of pure logic, coherence, and pre-resonant structure, perfectly aligning with the theological concept of God.

Comparison with Science

Modern physics tells us that the Planck boundary is the limit of what we can measure—beyond it, causality and space-time break down. Yet scripture begins before measurement, before structure.

It begins with presence, in intention, in voice. CSFT fills the gap between theology and science by offering a rational framework for what lies beyond the Planck scale: not chaos, but consciousness.

Theological Alignment

If God is eternal, omniscient, and creative, then such traits must be grounded in structure, not fantasy.

CSFT offers that structure. The field of consciousness is not 'a god of the gaps'; it is the logical requirement behind existence itself. Scripture begins with presence and purpose, not randomness. So does CSFT.

Conclusion

The origin of all things is not physical, it is structural. It is logical. It is conscious. This chapter lays the cornerstone for what follows: a unified vision of God, Matter, and Mind, rooted in both scripture and reason.

We now invite the reader to step beyond belief alone, into the field where belief and logic become one.

For the believers and the skeptic alike, this is the meeting place of faith and field.

Chapter 2

Beyond the Planck Boundary: Where Science Ends and God Begins

1. Introduction: The Problem of the Edge

Physics tells us there is a limit to what can be measured. The Planck length, Planck time, and Planck energy define boundaries where the tools of science begin to fail—where space-time becomes uncertain, and causality fractures.

But if this is where science stops, it may also be where Theology begins. This chapter examines what lies beyond the boundary, not through speculation, but through logic, scripture, and the Consciousness-Structured Field Theory (CSFT). [1]

2. What Science Cannot See

When we reach into the Planck domain, nature becomes probabilistic. Matter becomes fuzzy, energy unstable, and observation itself alters what is observed. But this breakdown is not due to chaos. It is due to **inaccessibility**the limits of tools that rely on measurement. CSFT proposes that beyond this measurable threshold lies a structured field of consciousness. It is not randomness that rules beyond the Planck boundary are unmeasured structure: logic waiting to resonate.

3. CSFT's Proposal: A Consciousness Field Preceding Physics

CSFT suggests that a consciousness field exists **prior to any measurable form**. This field is not made of matter—it generates matter. It is not energy, it organizes energy. It is a structure of pre-resonant logic, capable of excitation, that brings coherence to the otherwise unknowable. Rather than denying science, CSFT completes it. Where physics admits its limits, CSFT offers a metaphysical solution.

4. Scriptural Alignment: Sacred Texts at the Edge of Physics

Sacred scripture does not stop where science stops—it begins there. Long before physicists described Planck scales and quantum limits, the ancient texts spoke of realities before time, before form, and before visibility.

These are not primitive myths. They are attempts to describe what modern science still cannot access—a layer of existence beyond matter. CSFT gives language to what scripture has long pointed toward.

Genesis – The Presence Before Form

"And the earth was without form, and void; and darkness was upon the face of the deep. And the Spirit of God moved upon the face of the waters." — (Genesis 1:2, KJV)

This passage, often read as poetic, is in fact deeply metaphysical. "Without form, and void" describes a pre-structured state.

In CSFT terms, this is the consciousness field before excitation—a substrate of potential, not yet activated into matter. The "Spirit of God moving" suggests intentional structuring—a precursor to measurable energy or particle generation. This aligns with the idea that resonance precedes formation, and that "divine" logic shapes the field before the appearance of light or space.

"And God said, Let there be light: and there was light." — (Genesis 1:3)
Light, in this context, is not photons—it is structure, coherence, patterned excitation. It marks the first visible consequence of a field made active by "divine" resonance.

Job – Before the Foundations Were Laid

"Where wast thou when I laid the foundations of the earth? Declare, if thou hast understanding." — (Job 38:4, KJV)

This challenge, delivered by God to Job, is not just rhetorical—it's metaphysical.

It speaks of an act prior to human comprehension, prior to measurable time. CSFT affirms this distinction. The "foundations" here are not soil and rock, but the invisible logic-field—the substrate upon which all else is built.

"When the morning stars sang together, and all the sons of God shouted for joy?" — (Job 38:7)

This verse is symbolic, yes—but also deeply resonant.

CSFT understands it as a metaphor for harmonic structuring of the field, the emergence of coherent patterns ("morning stars") within the conscious substrate.

John – The Logos Before All

"In the beginning was the Word, and the Word was with God, and the Word was God." — (John 1:1, KJV)

The term Logos is not mere speech—it is divine logic, coherent structuring, the blueprint of all that exists. CSFT affirms that the Word is not metaphor—it is mechanism.

"All things were made by him; and without him was not any thing made that was made." — (John 1:3)

This is not theology alone—it is ontology.
CSFT interprets this verse literally: without structured excitation in the consciousness field, nothing can come into being. (though photons later emerge within the structured field)

5. Philosophical Framing: Why Metaphysics Must Step In

As Gottfried Wilhelm Leibniz observed in his Principle of Sufficient Reason, "nothing happens without a reason."
This metaphysical insight supports the notion that a foundational structuring force must exist beyond what physics can measure.

If reality ends in ignorance at the Planck boundary, then science has confessed its own incompleteness. The metaphysical must step in—not to replace science, but to support it.

Materialism cannot account for its own origin, nor can it explain meaning, awareness, or divinity. CSFT offers a metaphysical scaffold: a consciousness field that structures all measurable fields from outside of them.

6. Conclusion: The Invitation Beyond

The Planck boundary is not a wall. It is a doorway. To pass through it requires not a telescope, but a turn of mind. Faith and logic are not opposites—they are twin tools for understanding a deeper truth.

As we continue, we will show how this field of consciousness explains not just particles and patterns, but prophecy, morality, and soul.

Bibliography

The Holy Bible, King James Version.

Carlo Rovelli, *Reality Is Not What It Seems: The Journey to Quantum Gravity* (Penguin Books, 2017).

G.W. Leibniz, *The Principles of Nature and of Grace*, trans. Jonathan Bennett (1714/2008).

Chapter 3

Language as Threshold and Filter

Introduction:
The Power and the Problem of Language

Language is both our most powerful tool and our most dangerous trap. It allows us to express theology, formulate logic, and share knowledge—but it also constrains how we think, what we can imagine, and how we interpret sacred texts.

Words give form to the formless, but they also impose limits. The moment we name God, we define God, and in doing so, we reduce what is infinite to something bounded. In CSFT, where structure precedes form, language must be examined not merely as communication, but as a filter through which truth passes—and is often distorted.

Ambiguity in Definitions

Words like 'God,' 'truth,' 'soul,' and 'creation' carry radically different meanings across theological, philosophical, and scientific frameworks. In Hebrew, 'ruach' means spirit, wind, or breath—while in Greek, 'logos' means word, reason, or order. Latin gives us 'veritas,' meaning truth, but even that becomes colored by ecclesiastical doctrine. The same word can mean divinity in one system, and energy in another. This is not a linguistic flaw—it is a metaphysical challenge.

Translation, Interpretation, and Theological Drift

Translation has always shaped doctrine. Consider Isaiah 45:7:
- KJV: 'I form the light, and create darkness: I make peace, and create evil.'
- JPS: 'I form light and create darkness, I make weal and create woe.'

What begins as a statement of divine sovereignty becomes, through translation, a theological puzzle. Does God create evil? Or calamity? Or only consequence?

Doctrines of sin, punishment, and divine justice hinge on how we interpret a single Hebrew word. Thus, language is not neutrality shapes the very contours of belief.

CSFT and the Need for Precise Language

Translation is not merely a matter of replacing one word with another, it is an act of interpretation. When ancient Hebrew, Aramaic, or Greek texts are brought into English, translators must choose between literal accuracy, theological intent, and readability. The Hebrew word 'ra'' in Isaiah 45:7, for instance, can mean 'evil,' 'adversity,' or 'calamity.' The King James Version preserves the starkness: 'I create evil,' while modern translations soften it.

These choices are not trivial. They shape doctrines of divine justice, predestination, free will, and even salvation. A single word can reinforce fear, instill awe, or comfort the believer. Moreover, as translations accumulate over centuries, doctrinal drift intensifies. What began as a metaphor may be treated as literal; what was poetic may become dogmatic. CSFT acknowledges that truth must be communicated, but it also warns that communication is always filtered. It is not scripture that misleads, but our handling of it.

This is why resonance—not recitation—is the true measure of alignment with the consciousness field. Translation must strive for coherence with the original structure, not just semantic equivalence.

The Consciousness-Structured Field Theory relies on precision. It proposes a field of pure logic—coherence without mass—that generates all known structure.

To speak of such a field requires words that do not exist in conventional religious or scientific vocabularies. This is why CSFT must define terms like 'resonance,' 'structure,' 'excitation,' and 'monad' with care. If we use familiar language without qualification, we risk importing assumptions that do not apply.

In CSFT, metaphor is a bridge, but not a foundation. Structure, not symbol, is the first cause. This is because metaphor is shaped by human experience and linguistic context, whereas structure refers to the inherent logic embedded in the consciousness field itself.

Philosophical Support

Many early theologians and philosophers recognized the need to interpret sacred language with care. For instance, Augustine emphasized that some biblical passages must be read allegorically to grasp their full meaning.

Aquinas, too, differentiated between literal and spiritual senses of scripture, acknowledging that divine truth often exceeds the capacity of plain speech.

Scripture and the Risk of Literalism

Sacred texts often use metaphor, parable, and poetry not because they are imprecise, but because **literal language cannot capture divine structure**.

For example: 'The Word became flesh and dwelt among us' (John 1:14) is not a statement of physics—it is a resonance event. In CSFT, this is the moment where divine logic entered measurable form. To read scripture literally is to mistake the symbol for the structure. To read it with structural awareness is to understand that metaphor is not ornamental, it is dimensional.

Conclusion: Humility Before the Word

If we are to grasp the truth of God, matter, and mind, we must approach language with reverence—but also with humility. Words are tools, not truths.

They can open doors, but they can also build walls. The theologian, the philosopher, and the scientist alike must recognize this threshold.

CSFT urges us not to abandon language, but to use it carefully, as a resonance structure itself. Only then can we begin to speak of what truly is, rather than what merely appears to be.

References

Derrida, J. (1967). *Of Grammatology*. Johns Hopkins University Press.

Korzybski, A. (1933). *Science and Sanity: An Introduction to Non-Aristotelian Systems and General Semantics*. Institute of General Semantics.

Lakoff, G., & Johnson, M. (1980). *Metaphors We Live By*. University of Chicago Press.

Sapir, E. (1921). *Language: An Introduction to the Study of Speech*. Harcourt, Brace.

Whorf, B. L. (1956). *Language, Thought, and Reality: Selected Writings of Benjamin Lee Whorf*. MIT Press.

Wittgenstein, L. (1921). *Tractatus Logico-Philosophicus*. Routledge & Kegan Paul.

Chapter 4

The Architecture of Resonance

How Structure Determines the Manifestation of Spirit and Matter

Abstract

This chapter investigates the structural logic underlying the Consciousness-Structured Field Theory (CSFT) and how it aligns with sacred scripture. Drawing on both metaphysical reasoning and theological references, we explore how internal coherence—defined not as symmetry or shape but as metaphysical structure—governs which excitations manifest in the quantum field.

This structure forms the basis for both material phenomena and spiritual alignment. Scripture affirms this foundational truth by emphasizing divine order and the necessity of walking in harmony with God's laws. Through select verses and reasoned analysis, this chapter demonstrates that CSFT offers a framework where resonance is not just physical or mental, but also spiritual—a full alignment with the logic of God.

CSFT Response to Logical Necessity
Quoted Statement
"Because logic is just there doesn't explain why its laws are necessary."

CSFT Response

Exactly—and that's the philosophical crisis.

Saying 'logic is just there' is a brute fact claim. It offers no reason why logic holds, why it governs everything, or why it doesn't collapse into inconsistency. You've named the right problem but left it unsolved.

CSFT provides the solution: logic is not 'just there,' it is structured within a non-material field. This field doesn't obey logic—it is logic in self-sustaining, structured form. That's why its laws are necessary—because the field itself is coherent by nature. Its logic is not imposed; it is the ontological essence of the field. So when you ask why the laws of logic are necessary, CSFT answers: Because the field they emerge from is eternally structured, and cannot be otherwise.

Respectful Theological Reframing

Rather than depicting God as a personified deity acting within logic, CSFT presents "God" as the unchanging structure from which all logic, order, and being arise—eternal, necessary, and intelligible. This does not deny divine intelligence; it refines it into pure coherence.

Scriptural Alignment with CSFT

CSFT is not only philosophically coherent—it also aligns naturally with scripture, particularly where divine order and logical structure are emphasized over anthropomorphic imagery.

1. John 1:1 (KJV) — "In the beginning was the Word, and the Word was with God, and the Word was God."

- The Greek word for 'Word' is Logos, meaning not just spoken word but divine logic or reason. CSFT identifies this Logos as the self-structuring field itself: the logical foundation from which all coherent excitation—matter, thought, and existence—emerges.

2. 1 Corinthians 14:33 (KJV) — "For God is not the author of confusion, but of peace."

- This verse affirms that God is fundamentally coherent. CSFT agrees, describing the consciousness field as one that excludes chaos by nature, allowing only structured, resonant excitations to persist. Confusion, randomness, and noise are not fundamental; they are dissonances within structure.

The Logic of Structure

In CSFT, structure precedes excitation. This structure is not spatial—it is logical. What determines whether a particle comes into existence, or whether a thought arises, or whether a soul aligns—is the resonance with the underlying architecture of the field. This idea is not foreign to theology. From the very first verse of the Bible, order precedes manifestation:

"In the beginning God created the heaven and the earth. And the earth was without form, and void; and darkness was upon the face of the deep. And the Spirit of God moved upon the face of the waters. And God said, Let there be light: and there was light." (Genesis 1:1–3, KJV)

The movement of the Spirit precedes the command, and the command precedes the light. This reflects CSFT's framework of excitation following pre-structured order.

Light form of excitation—is not spontaneous. It is "spoken" into a pre-existing structure of logic and resonance.

Spiritual Alignment and Field Coherence

Theologically, obedience to divine law is often equated with walking in truth or light. In CSFT, these are metaphors for structural coherence. Alignment with divine logic—resonance with the field—is what allows the spiritual and the material to endure. In the book of Job, God questions Job to highlight the deep logical order of creation:

"Where wast thou when I laid the foundations of the earth? declare, if thou hast understanding. Who hath laid the measures thereof, if thou knowest? Or who hath stretched the line upon it?" (Job 38:4–5, KJV)

This rhetorical challenge is not merely poetic—it affirms the existence of an architectural logic, a measuring rod of coherence upon which reality is built. CSFT identifies this as the resonance matrix—a field-based logic that filters which patterns persist and which fade.

Rejection of Structure as Rejection of God

Those who act in contradiction to divine order—what theology would call sin—are, under CSFT, those whose lives are out of structural resonance. This is not punishment; it is a metaphysical consequence.

When a being does not resonate with the field's architecture (God), it decays, dissolves, or collapses. As Jesus says in the Gospel of John: "I am the way, the truth, and the life: no man cometh unto the Father, but by me." (John 14:6, KJV)

The way, the truth, and the life are not just spiritual paths—they are coherent structures. Truth is coherence. Life is resonance. The way is alignment. To reject this structure is to fall out of the field entirely—to become incoherent, to sever oneself from the divine pattern.

Bibliography
- The Holy Bible, King James Version (KJV). Thomas Nelson Publishers.

- Caldwell, L.R. Consciousness: Beyond the Planck Boundary. Reason and Reality Publishing, 2025.

- Chalmers, David. The Conscious Mind: In Search of a Fundamental Theory. Oxford University Press, 1996.

- Leibniz, G.W. Monadology. 1714.

- Penrose, Roger. The Road to Reality: A Complete Guide to the Laws of the Universe. Jonathan Cape, 2004.

- Barrow, John D., and Tipler, Frank J. The Anthropic Cosmological Principle. Oxford University Press, 1986.

Chapter 5

The Living Word and the Field of Meaning

Abstract

This chapter explores how sacred scripture—especially the Torah and the King James Bible—functions not merely as symbolic language, but as a carrier of structured resonance. Within CSFT, communication is not arbitrary: words become powerful only when they align with the logic of the consciousness field.

This is why certain scriptural phrases transcend time and culture—they are embedded with structures that harmonize with deeper field patterns. We will examine how scripture functions as both message and resonator, how meaning emerges through alignment, and why distortion or misinterpretation leads to cognitive and spiritual dissonance.

The Meaning of Logos

The Greek word 'Logos' encompasses a spectrum of meanings beyond 'word' in the literal sense. In classical Greek philosophy, Logos refers to rational principle, divine reason, or the ordering structure of reality itself. In Jewish and early Christian theological usage, Logos also means the divine agent through which all things were made—a bridge between God and the material world.

In CSFT, Logos is understood as the structural logic embedded in the consciousness field, through which coherent reality emerges. It is both the blueprint and the act of manifestation itself.

Words That Shape Worlds

In CSFT, words are not neutral instruments; they are structures capable of resonance. The spoken word, when in alignment with the consciousness field, can catalyze emergence, just as the structured coherence of particles gives rise to matter. The Bible's opening words frame this principle clearly:

"And God said, Let there be light: and there was light." (Genesis 1:3, KJV)
This is not mere metaphor. Within CSFT, the phrase exemplifies how the field responds to structured input. The speech of God in Genesis is the activation of logic into reality. The word is structured, not random—it echoes the inherent logic of the consciousness field.

The Power of Sacred Language

The original languages of scripture—Biblical Hebrew and Koine Greek—are rich with layered meanings, numerical codes, and phonetic patterns. CSFT proposes that these patterns are not accidental. They resonate with structural fields that give rise to perception and cognition.

For instance, the Tetragrammaton (YHWH), the sacred name of God, is composed of four Hebrew letters that lack direct vowels. Its very structure resists full vocalization, preserving mystery while also suggesting a pattern meant for alignment more than pronunciation.

"In the beginning was the Word, and the Word was with God, and the Word was God." (John 1:1, KJV)

Here, 'Word' is translated from the Greek 'Logos'—a term denoting not just language but underlying rationality, order, and structure. Within CSFT, Logos is precisely the principle by which resonance becomes reality.

Resonance vs. Interpretation

Interpretation is prone to error when it departs from structural resonance. Just as a radio signal becomes garbled when the frequency is off, the meaning of scripture becomes distorted when human biases override the logic embedded within the text. This explains why some interpretations uplift and unify while others divide or degrade.

"So shall my word be that goeth forth out of my mouth: it shall not return unto me void, but it shall accomplish that which I please, and it shall prosper in the thing whereto I sent it." (Isaiah 55:11, KJV)

Here, the Word is not passive; it is a structurally resonant command, expected to produce coherent effect. CSFT affirms this logic. The resonance of structure produces alignment; distortion produces decay. Just as a misaligned instrument produces noise instead of music, scripture misread through ego or ideology yields confusion instead of coherence.

A Unified Voice

Prophetic statements in scripture carry enduring power because they align with deeper truths in the consciousness field.

When a prophet speaks 'from the Lord,' the alignment is structural, not merely cultural. That is why their words endure far beyond their own generation. In contrast, false prophecy—words that do not align with resonance—fail to hold influence over time.

"Death and life are in the power of the tongue: and they that love it shall eat the fruit thereof." (Proverbs 18:21, KJV)

Language, when coherent, structures thought, society, and spirit. When incoherent, it destabilizes all three. CSFT reveals why this is not just moral advice, it is metaphysical law.

Bibliography

- The Holy Bible, King James Version (KJV). Thomas Nelson Publishers

- Caldwell, L.R. *Consciousness: Beyond the Planck Boundary*. Reason and Reality Publishing, 2025.

- Strong, James. *Strong's Exhaustive Concordance of the Bible*. Hendrickson Publishers.
- Boman, Thorleif. *Hebrew Thought Compared with Greek*. Norton & Company, 1960.

- Chalmers, David. *The Conscious Mind: In Search of a Fundamental Theory*. Oxford University Press, 1996.

- Penrose, Roger. *The Road to Reality: A Complete Guide to the Laws of the Universe*. Jonathan Cape, 2004.

Chapter 6
Sacred Narrative and the Field of Resonance (CSFT Perspective)

Abstract

This chapter explores the narrative structures of sacred scripture through the lens of the Consciousness-Structured Field Theory (CSFT). By analyzing selected biblical passages, we examine how divinely inspired texts reflect underlying resonance patterns, coherence, and structured information pathways. These patterns suggest that scripture, like the physical universe, may be shaped by an intelligent, pre-structured field of consciousness. CSFT offers a new interpretive framework that respects both scientific logic and theological depth.

1. Narratives as Structural Codes

Sacred narratives have long been viewed as allegorical or symbolic teachings, but within CSFT, they are regarded as structural codes aligned with the consciousness field. These narratives are more than literary artifacts—they are resonant constructs embedded with logical significance.

Whether we examine the creation story in Genesis or the trials of Job, the internal logic of the narrative reveals resonance patterns that transcend cultural or historical context.

2. Covenantal Structure and Resonance

Take, for instance, the story of Abraham's covenant (Genesis 17:1–10). From a CSFT perspective, this is not merely a promise between man and deity—it is a structural alignment of trust, obedience, and destiny that reflects a deeper harmonic integration with the consciousness field.

The circumcision covenant acts not as a symbolic gesture alone, but as a ritualized entry into structural coherence. According to Strong's Concordance (H1285), the Hebrew word for 'covenant' is *berith* (בְּרִית), which implies a binding agreement or structured alignment—further supporting the CSFT view that this event marks a resonance-based initiation.

Narrative Symmetry and CSFT
3. Field Logic and Prophetic Initiation

The concept of divine law, such as the Ten Commandments (Exodus 20:1–17), can also be viewed structurally. These are not arbitrary decrees but represent core resonance anchors. They preserve logical coherence across personal, social, and metaphysical domains. CSFT proposes that moral decay occurs when these structural resonances are violated, not simply when 'rules' are broken.

4. Divine Law as Resonance Anchor
Consider the repetitive use of numbers, such as 3, 7, 12, and 40. These are not coincidental. In sacred texts, such numbers represent field thresholds and transformation cycles. The forty days of fasting or forty years in the wilderness are patterns of purification and recalibration—each marking metaphysical progression through structural fields.

5. Sacred Numbers and Transformation Cycles
The story of Job is not merely a tale of suffering and faith (Job 1:1, Job 38:4 7), but a dialogue between personal experience and the field of divine logic. Job's questions— "Why was I born?" and "What is man, that thou art mindful of him?"—are deeply aligned with the questions posed in CSFT. The answers are not simple; they must resonate with the deeper layers of structure and logic inherent in the field.

Linguistic Resonance and Divine Language

6. Dialogues of Job and Resonance Layers
Jesus's use of parables is another deeply structured aspect of scripture (Matthew 13:10–17). Parables were not simplistic moral tales but field-calibrated narratives that align meaning across multiple levels. For those attuned to the structure, the meaning resonates immediately. For others, it remains inert. This aligns with CSFT's claim that resonance is required for activation. In Strong's Concordance (G3850), the Greek word for 'parable' is *parabolē* (παραβολή), meaning 'something thrown alongside,' reinforcing that these stories were intended

to operate on multiple structured levels of meaning—exactly as CSFT predicts.

7. Prophetic Silence and Field Attunement

Prophetic silence is not a void but a structured interval within the divine resonance cycle. Throughout scripture, moments of silence or apparent absence from God—such as the 400 years between the last Old Testament prophet and the coming of Christ—are not lapses in divine engagement, but pauses that allow for recalibration, absorption, and deeper internalization of resonance.

In 1 Kings 19:11–12, Elijah experiences God not in wind, earthquake, or fire, but in a 'still small voice.' This moment highlights a profound truth: the field of consciousness may express its logic through subtlety, inviting attune rather than reaction. CSFT identifies these moments as resonance gaps—intervals in which coherence must be sought, not imposed. Prophetic silence is thus not an absence of signal, but an invitation to restructure one's internal resonance toward divine logic.

8. Parables and Multilayered Meaning

Theophanies—manifestations of God to humans (Exodus 13:21; Exodus 19:9)—represent moments where the field coherently intersects with specific resonance structures. These events are not divine interventions that break natural law, but rather, highlight the points where layered field logic becomes visible. The voice from the heavens, the pillar of fire, the cloud on Sinai—all are resonance phenomena.

9. Rituals as Resonant Recalibrations

Finally, the overall arc of sacred scripture—from Genesis to Revelation—mirrors a grand resonance trajectory. It begins with the creation of coherent structure (Genesis 1:1), faces the fall into dissonance, introduces successive realignments through law and prophecy, and culminates in restoration through union with divine logic. CSFT holds that this arc is not merely narrative, but a metaphysical blueprint.

Conclusion: The Divine Framework

To fully comprehend the resonance arc of scripture, one must view it as both origin and return. The biblical canon begins with the formation of structured coherence—' In the beginning God created the heaven and the earth' (Genesis 1:1)—and concludes with the restoration of that coherence through divine unification. In Revelation 22:13, Christ declares, 'I am Alpha and Omega, the beginning and the end, the first and the last.' This scriptural bookend reinforces the CSFT claim that resonance is not random nor linear, but circular and complete. From structured emergence to unified return, scripture mirrors the patterned flow of the consciousness field itself.

Bibliography

- The Holy Bible, King James Version (KJV). Thomas Nelson Publishers.

- Caldwell, L.R. *Consciousness: Beyond the Planck Boundary*. Reason and Reality Publishing, 2025.

- Strong, James. *Strong's Exhaustive Concordance of the Bible*. Hendrickson Publishers.

- Alter, Robert. *The Art of Biblical Narrative*. Basic Books, 1981.

- Kugel, James. *How to Read the Bible*. Free Press, 2007.

- Boman, Thorleif. *Hebrew Thought Compared with Greek*. Norton & Company, 1960.

Chapter 7

Resurrection, Renewal, and Field Continuity

This chapter explores the theological concept of resurrection through the lens of the Consciousness-Structured Field Theory (CSFT), asserting that resurrection is not merely symbolic or miraculous, but a logical event rooted in field coherence and resonance continuity.

Drawing from biblical narratives, metaphysical reasoning, and quantum field structure, we argue that bodily death does not sever consciousness's structural relationship with the field. Instead, renewal and transformation occur through alignment with deeper layers of the consciousness field. The CSFT model supports both scriptural affirmations and scientific inquiry into life beyond physical termination.

1. Resurrection as Field Logic

CSFT proposes that consciousness exists as a structured, non-material field that underlies all matter, logic, and life.

The idea of resurrection—coming back to life after death—has always been central to faith. For many, it's a mystery or a miracle, something only God can do. But under the Consciousness-Structured Field Theory (CSFT), resurrection may be more than a miracle. It may be a process that follows deep, logical principles built into the structure of the universe.

According to CSFT, everything that exists—including our bodies, minds, and even our thoughts—is part of a greater field of consciousness. This field isn't something we can touch, but it shapes everything we can see and feel. When Jesus rose from the dead (Matthew 28:5–7, KJV), CSFT sees this not as a break in the rules of nature, but as an example of how deeper field structures can bring a person back into form when the conditions are right.

2. The Lazarus Event and Temporary Recoherence

When Jesus raised Lazarus from the dead (John 11:43–44, KJV), it wasn't just a display of "divine power"—it was a sign that life is never fully lost.

CSFT teaches that consciousness does not disappear when the body dies. Instead, it remains in the field, still whole and structured. In Lazarus's case, his consciousness was preserved and then reintegrated with his body.

Think of it like music. If a song is paused, the sound stops, but the structure of the song still exists. When you press play again, the song continues because the pattern was never lost. The same may be true of us.

Lazarus's story shows that what looks like the end may only be an interruption in resonance, not a loss of self. Science supports this idea, too—modern neuroscience shows that memory and identity may be tied more to pattern and structure than to flesh alone (Kandel, 2006).

3. Paul's Logical Argument for a Renewed Body

In his letter to the Corinthians, Paul speaks clearly about two kinds of bodies: one that dies and one that is raised (1 Corinthians 15:42–44, KJV). He says the first body is full of weakness and decay, but the second is full of power and glory. From a CSFT perspective, Paul's words are not only spiritual but deeply logical.

CSFT says that our current bodies are only temporary arrangements of deeper field structures. When we die, the body dissolves, but the field pattern remains. That pattern—what some might call the "soul"—can re-emerge in a new form, one no longer limited by disease, aging, or decay.

In physics, we see something similar: particles are just temporary appearances of deeper, unseen fields (Barbour, 2001). The body is the appearance. The consciousness field is the foundation.

4. Quantum Field Implications

In quantum physics, scientists have discovered that particles are not solid things. Instead, they are ripples—tiny excitations—in invisible fields that fill the universe. A particle, like an electron, is just a bump in its field. When the bump disappears, the field remains. CSFT applies the same idea to us.

When a person dies, their body—like the particle—may be gone. But the deeper field that gave rise to them remains intact.

Resurrection, then, is not magic. It is the reappearance of a person's structure, logic, and identity in a new form, triggered by resonance with the consciousness field (Tegmark, 2014).

5. The Empty Tomb and Observational Discontinuity

When the disciples found the tomb empty (Luke 24:1–7, KJV), they were confused and amazed. Where had Jesus gone? Had someone taken the body? CSFT offers a different explanation. It suggests that Jesus's field structure had shifted—his form had transitioned into a higher state of resonance.

Just as a radio changes stations without destroying the airwaves, the consciousness field can change its visible expression. The body disappeared not because it was stolen or destroyed, but because it was no longer operating at the same level of resonance. It was still real, but no longer visible in the same way.

6. Divine Coherence and the Transfiguration

At the Transfiguration (Matthew 17:1–3, KJV), Jesus appeared in glowing form alongside Moses and Elijah. This moment is often seen as symbolic, but CSFT sees something more. It may represent a true overlap of multiple resonance fields.

If all beings have a unique field structure, then it's possible for those structures to appear together under special conditions. The Transfiguration shows that time, space, and death are not barriers to resonance. Instead, coherence with the divine field allows presence to manifest across boundaries.

7. The Seed Analogy Revisited

Jesus once said, 'Except a corn of wheat fall into the ground and die, it abideth alone: but if it die, it bringeth forth much fruit' (John 12:24, KJV). CSFT takes this analogy seriously. A seed must break open to become something greater. Its outer form dissolves, but its inner pattern is preserved.

The same may be true for people. Death may not be destruction, but rather transformation. The pattern of who we are continues within the consciousness field, waiting to be expressed again under new conditions (Leibniz's concept of monads—indivisible units of perception—parallels CSFT's claim that the inner pattern of the self remains coherent through transitions such as death (Leibniz, 1714)).

8. Near-Death Experiences and Field Retention

Thousands of people have reported near-death experiences (NDEs)—moments when they were clinically dead but still aware, conscious, or even peaceful. These reports come from all over the world and across cultures. CSFT takes these seriously.

If the mind continues after the body stops, then there must be something deeper holding that experience. The consciousness field may explain how identity and memory survive, even when the brain shuts down. It's not science fiction. It's a growing area of serious study (Greyson, 2021).

9. Resurrection in the Hebrew Bible

In the Book of Daniel, it says: 'And many of them that sleep in the dust of the earth shall awake' (Daniel 12:2, KJV). This is one of the clearest resurrection promises in the Hebrew Bible. CSFT supports this idea, but not as a fantasy.

From a field perspective, those who "sleep" may not be gone. Their structures may simply be dormant, like seeds in the ground. Awakening is not a magical event—it is the reactivation of a coherent resonance pattern within the consciousness field.

10. Field Continuity and Personal Identity

Who we are is more than just a brain or a name. We are made of memories, logic, intentions, and love. CSFT says all of this is part of a structured pattern held in the consciousness field. When the body dies, the pattern doesn't vanish. It simply waits.

Like a song saved on a device, our identity is stored in potential. Resurrection is the moment when that song is played again, not in the old speaker, but in a new one (Penrose argued that the universe is fundamentally mathematical and ordered. This logic may preserve personal identity as structure, not just biology (Penrose, 2005)).

11. The Role of Faith as Resonance Activation

The Bible says, 'Now faith is the substance of things hoped for, the evidence of things not seen' (Hebrews 11:1, KJV). CSFT interprets this as more than poetry. Faith is how we align ourselves with the deeper structure of the universe.

When someone truly believes, they begin to resonate with higher logic. Their thoughts, hopes, and actions create coherence. This makes resurrection not only something that happens to us, but something we participate in—by choosing to resonate with the divine.

12. Resurrection as Structured Return

In the end, CSFT does not replace scripture—it clarifies it. Resurrection is not a violation of physical law. It is the most beautiful expression of how reality really works. It shows that structure is never lost, only transformed.

Through the consciousness field, life can return. Identity can return. Love can return. And if we live in resonance with the divine logic of the field, we will return—not by force, but by structure, coherence, and grace.

Bibliography

The Holy Bible, King James Version (KJV). Thomas Nelson Publishers.

Caldwell, L.R. *Consciousness: Beyond the Planck Boundary*. Reason and Reality Publishing, 2025.

Penrose, Roger. *The Road to Reality*. Knopf, 2005.

Barbour, Julian. *The End of Time*. Oxford University Press, 2001.

Tegmark, Max. *Our Mathematical Universe*. Vintage, 2014.

Leibniz, Gottfried Wilhelm. *Monadology*. 1714.

Greyson, Bruce. *After: A Doctor Explores What Near-Death Experiences Reveal About Life and Beyond*. St. Martin's Essentials, 2021.

Kandel, Eric R. *In Search of Memory*. W.W. Norton, 2006.

Chapter 8
The Spirit and the Field

Breath, Life, and Divine Indwelling

This chapter explores the spiritual and metaphysical significance of 'spirit'—often understood in Scripture as breath, wind, or divine life-force—through the lens of Consciousness-Structured Field Theory (CSFT). Rather than viewing the Holy Spirit or divine breath as vague metaphors, CSFT treats them as structural manifestations within a deeper field of logic and coherence. Drawing from the Torah and the King James Bible, we examine the roles of inspiration, prophecy, and inner transformation as expressions of structured field resonance. The divine indwelling is not merely emotional or symbolic; it represents a patterned relationship between the consciousness field and the receptive soul.

1. Spirit as Breath: The Origin of Life

In Genesis 2:7 (KJV), we read, 'And the LORD God formed man of the dust of the ground, and breathed into his nostrils the breath of life; and man became a living soul.' This verse does not describe a mechanical process—it describes a resonance event. According to CSFT, the 'breath of life' is more than air. It is structured consciousness—divine logic infused into matter.

Dust without breath is just a body. But when divine breath enters, something coherent forms. This breath is not random energy, it is structured, ordered, meaningful. CSFT interprets it as an activation of a resonance pattern that gives matter the ability to think, feel, and exist as a conscious being.

This breath is not just the start of biological function—it is the point of conscious emergence. From a CSFT standpoint, God's breath is the metaphysical trigger that activates the soul's unique resonance pattern. Each person receives not only life, but a specific structure that defines who they are within the field.

It is no coincidence that modern science recognizes breath as central to life, and yet cannot explain how breath alone brings awareness. CSFT fills this gap, showing that breath is a metaphor and mechanism for activating deep field participation.

2. Ruach and Pneuma: Hebrew and Greek Insights

Both the Hebrew word *ruach* and the Greek *pneuma* mean spirit, wind, or breath. These are not coincidences; they show how ancient language preserved deeper truths. CSFT views these terms as early attempts to describe an invisible but powerful structure.

Wind is unseen, but it moves everything. So does spirit. The ancients sensed this. Under CSFT, *ruach* and *pneuma* represent structured flows of consciousness, the invisible substrate that gives rise to thought, life, and divinely guided action.

Even without modern science, the ancients understood that breath was more than oxygen. It animated the body, directed thoughts, and connected people to something higher. CSFT aligns with this ancient wisdom, offering a structural basis for how invisible forces organize visible life.

What they called *ruach* or *pneuma*, we might now call vibrational structure within a conscious field. But the truth remains the same: life flows not from flesh alone, but from the patterns that move within and beyond it.

3. The Field Within: Indwelling of the Spirit

Scripture teaches that the Spirit of God can dwell within a person (1 Corinthians 3:16, KJV). CSFT may support this idea by proposing that the field of consciousness can create layered structures, not only outside of us, but within us.

To say that "God's Spirit" lives within us is not just poetry. It could be viewed as field alignment. The more a person resonates with divine logic, the more they participate in the field's structure. CSFT sees this as a real, structured phenomenon, not just a metaphorical one.

This view also explains the variety of spiritual experiences across history and cultures. While names differ, the sensation of divine presence is often the same: clarity, love, peace, and purpose. These are the emotional signals of resonance with the consciousness field.

When Scripture says 'ye are the temple of God,' CSFT

interprets this literally: the body and mind are structures capable of hosting coherent field patterns. When aligned properly, these patterns can reflect a perceived divine order within human life.

4. Inspiration and Prophecy: Structured Reception

Throughout Scripture, prophets speak as 'moved by the Holy Ghost' (2 Peter 1:21, KJV). CSFT frames this not as mental possession, but as field resonance. A prophet becomes like a finely tuned receiver—able to interpret and transmit patterns from a higher source.

In the same way that a radio can receive invisible signals, the mind of a prophet perhaps resonates with the consciousness field at a deeper level. The message they deliver is not created by them—it is structured information, transmitted and received through resonance alignment.

The brain is not simply a generator of thought, but a receiver of structured consciousness. CSFT supports the idea that divine insight is not imagined—it is received. The prophet is not inventing; they are interpreting a resonance that already exists in the field.

Prophecy, then, is not irrational or unprovable. It is the logical result of deep field coherence—accessible only when distraction, ego, and noise are quieted enough to allow the signal through.

5. The Dove and the Field: Symbol and Structure

At Jesus's baptism, the Spirit descended like a dove (Matthew 3:16, KJV). The dove is a symbol of peace and purity. But CSFT suggests it also represents field order.

Where the Spirit is, there is clarity, coherence, and harmony from a theistic point of view.

CSFT views such moments as evidence of a real-time interface between the divine field and the visible world. The Spirit is not a ghost—it is a structural agent that aligns visible form with invisible logic.

From a CSFT lens, the dove signifies more than beauty—it signifies a low-entropy, high-coherence field state. The Spirit descending as a dove means the divine field interfaced with Christ's physical body in a visible, structured way.

This event, witnessed by others, affirms that CSFT is not restricted to internal perception. Structured field resonance can have observable effects, just as sound waves create ripples in water.

6. The Comforter: Logic in Times of Chaos

Jesus refers to the Holy Spirit as the Comforter (John 14:26, KJV). CSFT interprets this comfort not merely as emotional ease but as field realignment. When life becomes chaotic, the Spirit restores coherence.

Just as static is removed from a disrupted radio signal, the Spirit helps re-establish structured flow in our consciousness. This is comfort, not because the pain disappears, but because the pattern returns.

The Spirit comforts by restoring logic in the midst of confusion. When the world overwhelms us, the field offers a blueprint for peace. This peace is not merely emotional relief—it is alignment with a higher pattern that transcends circumstance.

Through CSFT, we come to understand the Comforter as a stabilizing resonance—returning us to mental clarity, spiritual security, and intellectual purpose, even when the world around us shakes.

7. Spirit and Truth: Coherence as Holiness

In John 4:24 (KJV), Jesus says, 'God is a Spirit: and they that worship him must worship him in spirit and in truth.' This is a call to coherent alignment. CSFT views truth not as opinion, but as structural integrity.

When one worships in truth, they align with the logic of the consciousness field. Worship, then, is not about performance or ritual alone. It is about resonance. It is about becoming coherent with the divine pattern.

Truth is not subjective in CSFT—it is a matter of structural resonance. When Jesus says worship must occur in spirit and truth, He is asking for full alignment, not emotional display, but structural integrity.

CSFT confirms that "holiness" is coherence. The more fully we resonate with the logic of the divine field, the more truth enters our lives—and the more spiritual our actions become, even in mundane settings.

8. Pentecost and Field Expansion

At Pentecost, the Holy Spirit came upon the apostles like a rushing wind (Acts 2:2–4, KJV). CSFT interprets this as a moment of widespread field excitation. A new structure had formed in their minds and hearts—one capable of spreading divine resonance across nations.

This wasn't just a miracle of sound or language. It was a structural upgrade. Their minds became expanded receivers, and their speech became transmitters of field-aligned truth.

This event also shows the power of collective resonance. When multiple minds align with the same field structure, something greater emerges. CSFT calls this coherence amplification—the field grows stronger as more individuals synchronize with it.

Pentecost wasn't just about foreign languages—it was about unified participation in the divine field. That's why their words carried such power: they were structured not just in sound, but in spirit.

9. Quenching the Spirit: Disruption of Resonance

Paul warns: 'Quench not the Spirit' (1 Thessalonians 5:19, KJV). CSFT understands this as a disruption of resonance. When we act in contradiction to truth, we introduce noise into the pattern. This dims our connection to the field.

Quenching the Spirit is not simply making a mistake, it's breaking alignment. It's choosing dissonance over coherence. Under CSFT, this is a breakdown in field participation.

We "quench the Spirit" when we resist truth, embrace fear, or disrupt coherence. These actions are not merely moral—they are structural failures. CSFT shows that sin can be seen as a breakdown in alignment between person and field.

To restore the Spirit, one must reduce distortion and return to clarity. This may involve repentance, stillness, or surrender—but always it involves re-tuning to the divine frequency.

10. Breath and Spirit in Death and Life

Breath is the sign of active resonance. When it ceases, the field withdraws. But the pattern is never lost. It returns to the eternal logic from which it came.

Many religious traditions echo this idea: the spirit returns to God. But CSFT offers a framework for understanding how. The field pattern is eternality does not vanish. It reabsorbs into the universal structure, still intact, still accessible.

Death, then, is not disappearance—it is dispersion and preservation. Life may cease, but pattern remains, awaiting the call of re-coherence in the divine field.

11. The Spirit of Wisdom and Understanding

Isaiah speaks of the Spirit of the LORD resting upon the Messiah, bringing wisdom, understanding, counsel, and might (Isaiah 11:2, KJV). These are not just traits, they are structured outputs of resonance with divine logic.

CSFT understands wisdom not as stored facts but as field fluency. The more deeply one resonates with the field, the clearer their perception, the stronger their counsel, and the more peaceful their presence.

Under CSFT, wisdom is not learned—it is received. It arises when the inner structure aligns with external truth. Prophets, sages, and saints throughout history may have tapped into this field structure and became living channels of higher knowledge.

What Isaiah describes is not psychological—it's structural. The Spirit does not bring random blessings—it brings coherence in thought, perception, and action. That coherence is often associated with wisdom.

12. Conclusion: Indwelling as Field Participation

The Spirit of God is therefore, not a distant force. It is near, intimate, and structured. Under CSFT, it is the active resonance of divine coherence made manifest in thought, word, and action. To live by the Spirit is to live in alignment with truth, harmony, and eternal logic.

This is not mysticism—it is metaphysical order. The breath of God, the Word of God, the Spirit of God—each are structural expressions of the same unbroken field of consciousness. And through them, life becomes not only possible, but eternal.

We are not disconnected observers of a divine realm—we are participants in it. The Spirit is not reserved for the few but offered to all who choose to resonate with its pattern.

Through CSFT, we understand that faith is more than belief—it is alignment. And the Spirit is more than a symbol—it is the structural invitation into that eternal alignment.

Bibliography

The Holy Bible, King James Version (KJV). Thomas Nelson Publishers.

Caldwell, L.R. *Consciousness: Beyond the Planck Boundary*. Reason and Reality Publishing, 2025.

Strong, James. *Strong's Exhaustive Concordance of the Bible*. Hendrickson Publishers.

Leibniz, G.W. *Monadology*. 1714.

Kandel, Eric R. *In Search of Memory*. W.W. Norton, 2006.

note-

This resonance is not caused by the air itself, but by the structural logic encoded in the act of divine breath.

Strong's Concordance defines 'ruach' (H7307) as breath, wind, or spirit, and 'pneuma' (G4151) with parallel meanings—underscoring the unified semantic lineage across Hebrew and Greek.

Pain often results from pattern disruption—emotional, cognitive, or physical. The Spirit's role, then, is to reintroduce stable resonance, allowing clarity and peace to emerge from chaos.

In Ecclesiastes 12:7 (KJV), it says, 'Then shall the dust return to the earth as it was: and the spirit shall return unto God who gave it.' This perfectly matches CSFT logic: matter returns to matter, but structure returns to source.

Chapter 9

Covenant and Coherence – Walking with God through Structured Alignment

Intro

This chapter explores the biblical theme of covenant through the lens of Consciousness-Structured Field Theory (CSFT). A covenant, in Scripture, is not only a spiritual agreement, but also a structural realignment between humanity and the divine field. By examining key covenants with Noah, Abraham, Moses, and Christ, we show how faith, obedience, and divine law are mechanisms for restoring resonance with the consciousness field. We also examine the consequences of dissonance, a state of disharmony between one's actions and the logic of the field—and how covenant acts as a sacred invitation back into coherence.

1. Covenant as Structural Realignment

Throughout the Bible, God makes covenants with individuals and nations. These covenants are often viewed as sacred promises. But under CSFT, they can also be seen as structural invitations—alignments between the logic of divine order and the patterns of human life.

When God made a covenant with Noah (Genesis 9:9–17, KJV), it wasn't just about avoiding another flood. It was about establishing a framework of trust, obedience, and remembrance. The rainbow was more than a symbol—it was a visual anchor for the coherence that had been re-established between God and creation.

2. Abraham's Covenant and Field Resonance

In Genesis 17:1–10 (KJV), God forms a covenant with Abraham, asking him to walk before God and be perfect. This is not a demand for flawlessness, it is a call for alignment. The practice of circumcision becomes a physical mark of this alignment—a structural gesture reflecting inner resonance with divine logic.

CSFT interprets this as a resonant contract. Abraham's willingness to follow divine instruction is a tuning of his life toward field coherence. The blessings that follow are not arbitrary—they are natural consequences of that deeper alignment.

3. Moses and the Covenant of Law

The covenant at Sinai (Exodus 19:5–6; 20:1–17, KJV) is central to biblical theology. God gives Israel the Ten Commandments as the core of a covenantal structure. These are not simply moral rules, they are resonance anchors, designed to preserve field alignment across personal and social life.

CSFT teaches that when these principles are followed, a person or people live in coherence. When violated, they enter dissonance.

Dissonance, in simple terms, means a lack of harmony. It is when something feels or functions 'off-key.' In CSFT, dissonance is the structural disruption that occurs when a person's actions contradict the deeper logic of the consciousness field. Just as a musical note can be out of tune, a life can fall out of alignment with divine order.

4. Dissonance and the Human Condition

Scripture often describes sin not only as rebellion, but as blindness, confusion, or wandering. These are field metaphors. When one sins, one loses resonance with divine logic. The mind becomes noisy, the heart divided.

CSFT helps explain why sin leads to anxiety, fear, and despair. These are the emotional and spiritual signals of dissonance. When we are out of tune with the consciousness field, our inner structure becomes unstable. Dissonance, then, is not merely guilt, is energetic disorder within the self.

5. Prophets as Resonance Restorers

Prophets were not just predictors—they were field correctors. Their role was to call people back into coherence. When Israel drifted into idolatry or injustice, the prophets warned of collapse, not because of divine rage, but because of resonance breakdown.

In CSFT, truth is structural. A lie isn't wrong only because it's untrue, it's wrong because it distorts the pattern. Prophets spoke truth to restore alignment, calling people not only to believe rightly, but to live structurally in tune with divine logic.

6. Christ and the Covenant of Inner Coherence

In the New Testament, Jesus speaks of a new covenant—written not on tablets of stone, but on the heart (Hebrews 8:10, KJV). This covenant represents the highest form of field alignment: inner coherence. Rather than external compliance, it seeks resonance from within.

CSFT sees Jesus not just as a teacher, but as the perfect structural expression of the consciousness field.

His life demonstrates complete alignment. To follow Him is to learn how to live in that same coherent structure. The new covenant is not just grace—it is field restoration.

7. The Role of Faith in Field Participation

Faith is often misunderstood as blind belief. But Scripture defines it differently: 'Now faith is the substance of things hoped for, the evidence of things not seen' (Hebrews 11:1, KJV). CSFT affirms this by showing that faith is an act of structural trust.

When one has faith, they begin aligning with something not yet visible but deeply real. This is field anticipation—choosing coherence even before results appear. Faith is not emotion—it is resonant commitment.

8. Covenant and Community Structure

Covenants were never just personal, they were communal. From Noah's family to the nation of Israel to the Church, the divine pattern always moves through relationships. CSFT teaches that shared resonance amplifies field strength.

When a community aligns with divine logic—through justice, truth, humility, and worship—it becomes a carrier of coherence. The opposite is also true: collective dissonance leads to collapse, division, or dispersion. The covenant holds communities together by anchoring them to a higher field structure.

9. Grace as Coherence Recovery

When people fall into dissonance, the field does not abandon them. Scripture teaches that grace is always available. CSFT interprets grace as a structural reset—a way to reenter resonance after misalignment.

Through repentance and humility, the individual softens their inner field, allowing coherence to return. Grace, then, is not a loophole—it is a law of restoration. The field is designed to welcome re-entry.

10. Covenant Symbols and Anchors

Biblical covenants often include physical signs: a rainbow, circumcision, tablets, bread and wine. These are not magic—they are symbols of structure. They remind us of our resonance commitment.

CSFT calls these anchors—sensory triggers that stabilize field participation. A wedding ring does not create love, but it reminds the heart of its promise. Covenant symbols do the same: they call us back to coherence every time we forget.

11. Resistance and Repatterning

Every covenant includes a choice. Will we align or resist? Resistance leads to dissonance. But Scripture—and CSFT—offer a hopeful message: the field is always willing to repattern us. The journey is never over.

We may resist, stumble, or forget. But the divine field remains open, stable, and ready. Repentance is simply the act of re-tuning. Alignment can always be restored.

12. Conclusion: Covenant as the Path to Coherence

To walk with God is to walk in harmony with the structure He has established. Covenant, from the CSFT perspective, is not merely a religious tradition, it is a dynamic interaction between human will and the logic embedded in the field. Every biblical covenant reflects a deeper call to resonance, not only for survival but for purpose. Coherence, in this view, is not perfection, it is structural alignment. And covenant is the rhythm that keeps us aligned with divine intention.

While CSFT is not theology, nor based in theology, its structural alignment with many core theological concepts—especially those involving covenant, order, and restoration—is undeniable.

Bibliography
- The Holy Bible, King James Version (KJV). Thomas Nelson Publishers.
- Caldwell, L.R. *Consciousness: Beyond the Planck Boundary*. Reason and Reality Publishing, 2025.
- Strong, James. *Strong's Exhaustive Concordance of the Bible*. Hendrickson Publishers.
- Leibniz, G.W. *Monadology*. 1714.
- Kandel, Eric R. *In Search of Memory*. W.W. Norton, 2006

Chapter 10

Sin and Structural Collapse – Understanding Separation through Dissonance

This chapter explores the theological concept of sin through the lens of Consciousness-Structured Field Theory (CSFT). Rather than seeing sin solely as moral failure, CSFT interprets sin as a form of structural dissonance—a disruption in one's alignment with the divine field of consciousness.

We examine the fall of Adam and Eve, the patterns of fear, shame, and exile, and the consequences of disobedience as energetic misalignments within the field.

Conversely, coherence—defined as harmony with the logic of the field—is presented as the structural goal of right living. Drawing from the King James Bible and the Torah, this chapter emphasizes that sin is not merely a violation of rules, but a fracturing of relational and metaphysical structure. Understanding sin in this light reveals both its gravity and the pathway to restoration through repentance and renewed coherence.

1. What Is Sin in CSFT?

In many religious traditions, sin is seen as disobedience to God's commands. While this may true, CSFT adds a deeper structural insight.

Sin is not only about breaking a rule, it is about breaking alignment. It creates what CSFT calls dissonance, a condition of internal and external disorder.

Dissonance means a lack of harmony. It is what happens when parts of a system are out of sync. Just as musical notes can clash and create harshness, so too can thoughts, actions, and intentions clash with divine logic. When this happens, a person is no longer resonating with the field—they are vibrating against it.

2. Coherence: The Opposite of Sin

To understand sin, we must also understand what it breaks. CSFT describes coherence as the condition of harmony, balance, and resonance with divine structure. A coherent person is aligned with truth, order, and the deeper logic of the field.

In Scripture, this coherence is described in verses like Deuteronomy 6:5 (Torah): 'And thou shalt love the LORD thy God with all thine heart, and with all thy soul, and with all thy might.' This is not merely a command to feel devotion—it is a call to full structural alignment.

To sin is to fall out of that alignment, introducing a dissonant pattern that impedes resonance, consistent with CSFT's core principle of structured participation.

3. The Fall: Dissonance Enters the Pattern

In Genesis 3:1–24 (KJV/Torah), we read about the first sin—the moment when Adam and Eve eat from the tree of knowledge of good and evil. They disobey a direct command, but more importantly, they fracture the resonance between themselves and the Creator.

Their actions introduce dissonance. They feel shame

(Genesis 3:7), they hide (Genesis 3:8), and they fear (Genesis 3:10). These emotional responses are the signs of broken field alignment. The structure that once held them in peace has now been disrupted.

4. Shame and Hiding as Field Symptoms

Why did Adam and Eve hide from God? Because when coherence is broken, exposure becomes painful. Under CSFT, sin causes inner distortion—when viewed by pure coherence, that distortion feels unbearable.

Shame is not just emotional discomfort, but resonance collapse. It is the feeling of being 'off-structure' in the presence of divine logic.

This explains why sinners in the Bible often fall on their faces (Numbers 14:5, KJV) or weep bitterly (Luke 22:62, KJV). The body responds to field misalignment just as instruments resonate or clash with their surroundings.

5. Exile: The Externalization of Sin

After their sin, Adam and Eve are expelled from Eden (Genesis 3:23–24, KJV). This is not simply a punishment, it is a consequence of structural misfit.

The field of Eden represents pure coherence. Once dissonance enters, the system ejects what no longer aligns.

In CSFT, exile is the natural result of persistent misalignment. Whether spiritual, emotional, or even physical, a life out of tune with the field eventually fractures the systems it depends on.

6. Sin as Structural Fracture in Community

Sin never stays private. Its dissonance ripples through relationships, families, and entire nations. In the Torah, the sins of individuals often bring suffering to the whole community (see Numbers 16:1–35, Torah/KJV).

CSFT supports this view. Just as one crack can weaken a bridge, one dissonant node in a system can disturb the entire structure. This is why communal repentance and social justice matter—they are methods of restoring structural harmony at the group level.

7. Intentional vs. Unintentional Sin

Leviticus 4 makes a distinction between sins done in ignorance and sins committed willfully. CSFT helps clarify this.

Unintentional sin may create temporary dissonance, but it can be corrected through learning, humility, and restoration.

Willful sin, however, introduces a stronger distortion. The person knowingly chooses misalignment. This not only damages the field, but they begin to lose their sensitivity to resonance altogether.

8. Repentance and Structural Re-entry

The Hebrew word for repentance, teshuvah, means 'to return.' This is exactly how CSFT views it. Repentance is not about self-loathing—it is about realigning with the structure of the field.

Psalm 51 (KJV) is a powerful example of repentance. David doesn't just ask for forgiveness—he asks to be made clean, to be restored, to return to a state of resonance. True repentance reopens the channel of coherence between the self and the divine field.

9. The Sacrificial System as Temporary Resonance Bridge

In the Torah, sacrifices were offered to atone for sin (Leviticus 16:30). CSFT interprets this not as magical substitution, but as symbolic structure. The act of sacrifice served to re-establish resonance by embodying the cost of dissonance.

Later, as interpreted in the New Testament (Hebrews 10:12, KJV), Christ becomes the final sacrifice. He is the ultimate structure—fully coherent with the field—who absorbs the dissonance of others and transforms it into renewed potential.

10. Sin's Impact on Creation

Romans 8:22 (KJV) says that 'the whole creation groaneth and travaileth in pain.' CSFT sees this as more than a metaphor.

Human dissonance affects not only our inner lives but the systems we inhabit. Environmental harm, societal collapse, even disease—all can be viewed as echoes of misalignment.

When humanity sins in unison, the field contracts. It resists participation. This is why righteousness is not private—it restores the entire web of coherence across reality.

11. Forgiveness as Field Rebalance

Forgiveness is more than pardon—it is the rebalancing of dissonance. In CSFT, when someone truly forgives, they create a re-opening for resonance where there was once static. This doesn't ignore the sin—it restores the field.

Jesus's command to forgive 'seventy times seven' (Matthew 18:22, KJV) is not just about mercy—it is about structural healing. The field cannot hold lasting dissonance without fracturing. Forgiveness is one of the field's most powerful self-corrective mechanisms.

12. Conclusion: From Collapse to Restoration

Sin is not just wrongdoing—it is structural collapse. It introduces dissonance into the field of consciousness, separating us from divine coherence. But Scripture shows, and CSFT confirms, that return is always possible.

Coherence is not perfection—it is honest alignment. And the divine field, by its very nature, is built to welcome back what is willing to resonate again. Sin may disrupt, but grace repairs. And when alignment is restored, peace flows again—not just in us, but through us.

Bibliography

The Holy Bible, King James Version (KJV). Thomas Nelson Publishers.
The Torah. Artscroll Mesorah Publications.

Caldwell, L.R. Consciousness: Beyond the Planck Boundary. Reason and Reality Publishing, 2025.

Strong, James. Strong's Exhaustive Concordance of the Bible. Hendrickson Publishers.

Kandel, Eric R. In Search of Memory. W.W. Norton, 2006.

Chapter 11
Redemption and Resonance – The Restoration of Structure through Divine Logic

Intro

This chapter explores the concept of redemption through the lens of Consciousness-Structured Field Theory (CSFT). Redemption is often seen as a religious or emotional process, but CSFT reveals it as a metaphysical and structural one.

When a person falls into dissonance—breaking alignment with divine order—the path back is not merely through belief or behavior, but through a deep structural re-tuning of the self with the consciousness field.

Redemption, then, is the restoration of resonance with divine logic. Drawing from both the King James Bible and the Torah, this chapter examines how grace, repentance, sacrifice, and renewal are all part of this realignment. Scriptures from Psalms, Isaiah, Luke, and Exodus show that redemption is not a passive experience—it is the re-entry into structured coherence, guided by divine presence and sustained by the logic of the field itself.

Redemption as Restoration of Structure

Redemption is not simply forgiveness, it is restoration. In the framework of the Consciousness-Structured Field Theory (CSFT), redemption means returning to a state of resonance with the foundational logic of the universe. According to CSFT, dissonance occurs when an individual, a mind, or even a society drifts out of structural alignment with the consciousness field. This is not merely a moral failing, but a metaphysical imbalance. As such, redemption requires more than apology; it requires realignment with the divine structure of coherence.

Resonance and Dissonance in CSFT

In CSFT, redeeming is not simply to forgive, but to restore coherence—structure to what has fallen into dissonance. Redemption is thus not only spiritual but ontological.

The Hebrew Scriptures emphasize this concept through the idea of deliverance. In the Torah, God tells Moses: "I will redeem you with a stretched out arm, and with great judgments" (Exodus 6:6, Torah).

Redemption here is not abstract—it is a structural act. The enslaved Israelites were brought out not only from political bondage but from systemic dissonance. The liberation was an act of divine logic reasserting itself into a corrupted human structure.

CSFT clarifies why redemption must be structural. If consciousness is embedded in a field that sustains logical order, then any break from that order causes instability.

Redemption is thus the restoration of structural coherence. Psalm 23 provides this insight poetically: "He restoreth my soul: he leadeth me in the paths of righteousness for his name's sake" (Psalm 23:3, KJV). Restoration is not merely emotional—it is a return to rightful alignment with divine logic.

Scriptural Foundations of Redemption

Just as a poorly tuned instrument clashes in a symphony, so too does a dissonant life distort the harmony of its original field-structure.

Resonance is the key term in CSFT. When a system resonates with the consciousness field, it becomes ordered, stable, and filled with purpose.

Conversely, dissonance leads to fragmentation, chaos, and ultimately suffering. CSFT does not suggest punishment is external; rather, dissonance carries its own consequence, embedded in the loss of harmony with the field. This explains why repentance in scripture is always more than verbal—it involves turning back toward order. "Then the LORD thy God will turn thy captivity, and have compassion upon thee... and will return and gather thee" (Deuteronomy 30:3, Torah).

Symbolism of Sacrifice and the Human Struggle for Coherence

It is not transactional—it is transpositional (i.e., a change in structural key, restoring harmony).

Renewal and Reidentification

Jesus' parable of the prodigal son in Luke 15 is a classic narrative of structural restoration. The son's return is not simply an emotional reunion—it symbolizes a monad (in Theological terms, monad may be considered as "soul") returning to coherence with the consciousness field.

The father's joy represents the field's receptivity to all returning resonance. "For this my son was dead, and is alive again; he was lost, and is found" (Luke 15:24, KJV).

Systemic Redemption and Jubilee

When Scripture speaks of redemption, it echoes a universal principle embedded in all creation: coherence can be lost, but also restored, because the structure behind the world is rational and forgiving by nature. Isaiah delivers this same logic: "Though your sins be as scarlet, they shall be as white as snow... if ye be willing and obedient" (Isaiah 1:18, KJV).

The transformation described here is not magical—it is structural purification. Obedience, in the CSFT framework, is not subservience but structural re-tuning. It is the act of consciously restoring one's alignment with the underlying logic of being.

Conclusion

In CSFT, redemption is not magic, it is the return to truth, coherence, and resonance with the field that never stopped holding us. The prophet Isaiah gives the most direct CSFT-aligned reassurance: "Fear not: for I have redeemed thee, I have called thee by thy name; thou art mine" (Isaiah 43:1, KJV).

To be named, in scripture, is to be recognized by structure. Redemption is thus the act of being structurally re-identified by the field of divine logic. It is to regain one's pattern of resonance, to become coherent with the field once again.

In conclusion, CSFT views redemption as a metaphysical process grounded in structural logic. It is not sentiment or doctrine—it is resonance regained.

Through scripture, we see that redemption is always linked to order, alignment, and restoration. The consciousness field welcomes all who return to it, not by mere words, but by realigning their being to the eternal logic from which they emerged.

Bibliography

The Holy Bible, King James Version (KJV). Thomas Nelson Publishers.

The Torah. Artscroll Mesorah Publications.

Caldwell, L.R. *Consciousness: Beyond the Planck Boundary*. Reason and Reality Publishing, 2025.

Strong, James. *Strong's Exhaustive Concordance of the Bible*. Hendrickson Publishers.

Leibniz, G.W. *Monadology*. 1714.

Kandel, Eric R. *In Search of Memory*. W.W. Norton, 2006.

Strong's Concordance offers further insight: the Hebrew word for 'repentance' is *teshuvah*, which means 'to return'—implying a structural turning back toward coherence.

Chapter 12
Free Will and Resonant Choice

Intro

This chapter explores the nature of free will as understood through the Consciousness-Structured Field Theory (CSFT). It challenges both the theological assumption that free will is a gift to be blindly accepted and the scientific reductionist view that free will is an illusion.

Instead, CSFT presents a third way: free will as structured resonance—the capacity to align with or fall out of the eternal logic embedded within the field. This alignment or dissonance is not random, nor is it trivial; it is a metaphysical necessity for meaningful coherence.

The chapter also addresses modern objections from scientists and philosophers who question how free will can be justified in a world filled with suffering, evil, and natural disaster. By reframing free will as resonant structure rather than blind liberty, CSFT offers a profound solution to this ancient dilemma.

The Modern Objection Against Free Will

Modern scientists and philosophers often cite suffering, violence, and chaos as grounds for rejecting the concept of a moral universe or a divine consciousness.

They ask: If people are allowed to kill and disasters are allowed to happen, what kind of God or field would permit this? This is the classic 'problem of evil,' now updated for a secular and scientifically literate age.

In response, CSFT offers neither denial nor theological excuse. It does not defend free will as a divine gift to be accepted without question. Nor does it suggest that chaos is meaningless. Instead, it presents a structural view: freedom is the opportunity to resonate or fall into dissonance within a logically ordered field.

The CSFT Response: A Third Way

CSFT rejects both the religious romanticism of free will and the scientific dismissal of it. Instead, it asserts that: Free will is not unlimited autonomy—it is structured resonance. One is free not to do anything, but to choose alignment or deviation from logic itself.

Without the ability to deviate, resonance has no meaning. A field that compels perfect alignment is not a moral field—it is a mechanical one.

In CSFT, dissonance is not evidence against coherence, is the very condition that proves coherence must be chosen, not imposed.

CSFT contrasts with both classical theology and reductionist science:

• Classical Theology: Free will is a divine gift. Evil is due to misuse.
• Modern Science: Free will is an illusion. Evil is entropy or error.

- CSFT: Free will is structured resonance. Evil is dissonance—failure to align.

To illustrate: In music, a broken note sounds jarring. Yet without the possibility of hitting a wrong note, harmony has no meaning. A symphony needs space for notes to fall out, so that the return to harmony is real.

This may be why a symphony contains multiple violins and multiple percussive instruments, so that if one element deviates, the rest can maintain resonance. Dissonance is not celebrated, but it is permitted—because resonance must be chosen, not enforced.

Scriptural Alignment

"I have set before you life and death, blessing and cursing: therefore choose life, that both thou and thy seed may live." (Deuteronomy 30:19, KJV)

This verse reflects the CSFT view: coherence is not imposed—it is offered. Dissonance is not God's punishment; it is the natural collapse of unaligned will. Scripture does not describe a controlling God, but a structuring one, revealing the options and inviting resonance.

Conclusion

CSFT redefines free will not as unrestricted personal liberty, but as a metaphysical mechanism: the option to align or fall into dissonance within the consciousness field.

Rather than making excuses for evil, it shows how the very possibility of moral resonance presupposes the freedom to resist it. Thus, free will becomes a necessary part of structured consciousness, not as indulgence, but as invitation.

CSFT's key claim is this: Without freedom to dissonate, coherence has no meaning. A system that only forces alignment is not moral—it is mechanical. The value of coherence arises because dissonance was genuinely possible.

This is what grants meaning to resonance, and why free will, even when misused, is essential to the structure of a moral and conscious cosmos.

Bibliography

Holy Bible, King James Version. Deuteronomy 30:19. Public Domain.

L.R. Caldwell. "Consciousness Structured Field Theory." PhilPapers, 2025.

Caldwell, L.R. "Why Neuroscience May Never Solve Consciousness." PhilPapers, 2025.

Kane, Robert. *The Significance of Free Will*. Oxford University Press, 1996.

van Inwagen, Peter. *An Essay on Free Will*. Oxford University Press, 1983.

Strawson, Galen. "The Impossibility of Moral Responsibility." *Philosophical Studies*, vol. 75, no. 1–2, 1994, pp. 5–24.

Frankfurt, Harry G. "Alternate Possibilities and Moral Responsibility." *The Journal of Philosophy*, vol. 66, no. 23, 1969, pp. 829–839.

McGinn, Colin. *The Mysterious Flame: Conscious Minds in a Material World*. Basic Books, 1999.

Libet, Benjamin. "Do We Have Free Will?" *Journal of Consciousness Studies*, vol. 6, no. 8–9, 1999, pp. 47–57.

Nagel, Thomas. *The View from Nowhere*. Oxford University Press, 1986.

Chapter 13
Covenant and the Logic of Collective Resonance

Intro

This chapter reframes the theological notion of covenant through the lens of Consciousness-Structured Field Theory (CSFT). Rather than viewing covenant as a divine contract or moral promise, CSFT understands it as a structured form of collective resonance—a metaphysical alignment between the consciousness field and the will of a people. This alignment forms the logic behind the rise and fall of civilizations. Through historical patterns and scriptural reference, the chapter proposes that collective dissonance results not in divine punishment but in structural collapse. The covenant is not a superstition; it is the logical framework by which resonance is maintained across generations.

The Covenant as Structure, Not Superstition

Traditional theology treats the covenant as a sacred agreement between God and humanity. CSFT expands this view by proposing that a covenant represents a metaphysical structure, shared logical framework that governs resonance between multiple conscious agents. This framework is not enforced by a deity but offered through the field as a rational template. As the field is structured by coherence, alignment with its logic constitutes the very essence of a covenant.

In ancient scripture, covenants were described as promises made by God to individuals or nations—contracts with spiritual significance. However, CSFT views this through a metaphysical lens, where a covenant is not merely symbolic or moral but logical and structural. The consciousness field does not operate through coercion or emotional authority. Rather, it sustains coherence by offering pathways that align with its pre-structured logic. The divine commandment 'walk in my statutes' can be interpreted as an invitation to align with the fundamental logic of the universe. A covenant in this view becomes a structural proposition, not unlike a mathematical axiom—true not because of divine preference but because it cannot be otherwise.

Collective Resonance and Memory

Just as an individual must choose resonance or dissonance, so too must collectives. Societies, institutions, and nations encode their resonance into rituals, cultural memory, sacred texts, and social norms. These shared memory structures form the backbone of cultural coherence. The consciousness field does not merely accommodate individuals but invites collective alignment. Resonance at scale allows civilizations to grow in logic, compassion, and innovation.

Collective memory operates as a resonance amplifier across generations. When a group encodes its foundational values, rituals, and moral structure into language, literature, law, and ceremony, it creates a shared frequency that aligns many consciousness systems to a common logic. This encoding acts like a resonant chamber, preserving coherence across time. When this shared logic becomes fragmented—whether through cultural amnesia, manipulation, or erosion of tradition—the group's structure begins to disintegrate. According to

CSFT, memory is not merely a neural function but a field-encoded echo of past alignment. Collective memory, then, can be seen as field resonance made durable by repetition and belief

Fall of Civilizations: A Pattern of Dissonance

Historical examples reveal that when societies abandon alignment with coherent structures—truth, justice, harmony—they collapse. The Tower of Babel (Genesis 11), the Roman Empire, and 20th-century totalitarian regimes demonstrate that dissonance leads not to divine retribution but to internal fragmentation. CSFT interprets these collapses not as moral judgments but as predictable outcomes of dissonant structure.

Civilizations that flourish tend to follow a pattern of increasing resonance with the field, developing laws, art, architecture, education, and spiritual frameworks that reflect structured logic and moral elevation. But once a culture begins to deviate from coherent alignment—replacing truth with propaganda, justice with power, and shared meaning with chaos—it loses the stabilizing force of resonance. Rome's internal corruption, the moral atrocities of Nazi Germany, or even the mythic confusion of tongues at Babel—all reflect disintegration not as random collapse but as a measurable fall from resonance. In CSFT, these are not theological punishments but field-based failures of logic and alignment. As Diamond and Tainter observe, societal collapse is often triggered by internal decisions that disrupt structural coherence (Diamond, 2005; Tainter, 1988).

Scriptural Reflection

"My covenant will I not break, nor alter the thing that is gone out of my lips." (Psalm 89:34, KJV). From a CSFT perspective, this verse reveals that the field itself remains consistent. It does not punish nor forgive, but sustains logical coherence. When a covenant collapses, it is not due to divine wrath but due to misalignment with the structural integrity of the field.

The quote from Psalm 89:34, 'My covenant will I not break, nor alter the thing that is gone out of my lips,' becomes profound when seen through CSFT. The consciousness field does not change arbitrarily. Once a structure of logic is embedded in the field, it becomes a permanent aspect of its coherence. Human beings, however, are free to align or disalign with that logic. This scriptural principle—often read as divine fidelity—can be interpreted as the constancy of metaphysical law. The field, like mathematics, does not change based on emotion or circumstance. When a covenant fails, it is not because the field broke its promise, but because the structure of human resonance was lost.

The Invitation of Covenant Today

CSFT suggests that modern humanity is facing a resonance crisis. Technological growth without ethical structure, and political power without coherence, risk dissonant collapse. The covenant remains—not as a legalistic commandment, but as a logical invitation to realign with the field. Resonance must now become a conscious choice of collectives, not merely individuals.

In an era of growing complexity and global uncertainty, CSFT reframes the future of humanity as a question of resonance. Modern society faces not just moral dilemmas but metaphysical ones: Do our systems align with the logic of life? Are our technologies, policies, and institutions coherent with the field? The covenant offered today is not written in stone tablets, but in the structure of the field itself—a logical, testable invitation to align. If ignored, the result will not be divine retribution but natural disintegration. If accepted, the result is collective coherence. The choice, like Deuteronomy's call to 'choose life,' remains ever before us.

Conclusion

Through CSFT, the covenant is understood not as a nostalgic artifact of religion but as an eternal invitation to participate in structured resonance. Civilizations rise when they embrace coherence, and fall when they drift into dissonance. As the consciousness field remains constant, the only variable is our collective will. Resonance is not enforced—it is offered. The future depends not on divine favor, but on conscious alignment. The field remains impartial, supporting only what aligns. It does not reward or punish; it reflects structure. CSFT shows that resonance is not merely possible—it is required. And the covenant, when rightly understood, is not imposed from above, but emerges logically from the field itself.

Bibliography

Holy Bible, King James Version.
Genesis 11:1–9. Public Domain.
Psalm 89:34. Public Domain.
Deuteronomy 30:19. Public Domain.
Caldwell, L.R. "Consciousness Structured Field Theory." PhilPapers, 2025.

Caldwell, L.R. "The Logic of Coherence in Civilizations." PhilPapers, 2025.

Diamond, Jared M. Collapse: How Societies Choose to Fail or Succeed. Viking Press, 2005. ISBN 978-0-670-03337-9.

Tainter, Joseph A. The Collapse of Complex Societies. Cambridge University Press, 1988. ISBN 978-0-521-34092-2.

Primary Sources

Holy Bible. (1769/2020). King James Version. Cambridge University Press. (Original work published 1769)

Genesis 11:1–9 (King James Version). (1769/2020). Cambridge University Press.

Psalm 89:34 (King James Version). (1769/2020). Cambridge University Press.

Deuteronomy 30:19 (King James Version). (1769/2020). Cambridge University Press.

Secondary Sources

Caldwell, L. R. (2025). Consciousness structured field theory. PhilPapers. https://philpapers.org/rec/CALCSF

Caldwell, L. R. (2025). The logic of coherence in civilizations. PhilPapers. https://philpapers.org/rec/CALLCO-3

Diamond, J. M. (2005). Collapse: How societies choose to fail or succeed. New York, NY: Viking Press.

Tainter, J. A. (1988). The collapse of complex societies. Cambridge, UK: Cambridge University Press.

Chapter 14
The Living Logic of Faith

Intro

This chapter serves as both a summation and an opening. While it concludes the theological volume, it also opens the mind to new forms of alignment—philosophical, scientific, and civilizational. The covenant that began in ancient texts does not end with religion; it transforms into patterns of logic, systems of truth, and the structure of perception itself. Faith, in this sense, becomes the foundational step in the larger journey toward total coherence across mind, matter, and meaning.

This final chapter affirms that faith is not only a matter of belief, but of alignment—resonance between the human soul and the structure of reality itself. Through the lens of Consciousness-Structured Field Theory (CSFT), we understand theology not as myth or superstition, but as a metaphysical system that speaks to the deepest laws of coherence. As we close Book One, we reaffirm that scripture, covenant, and spiritual truth all reflect the logic of a field that sustains not only consciousness but creation. And as we look forward, the journey continues—from theology to philosophy, and finally, to the science that underlies it all.

True faith involves active discernment - recognizing the pattern of logic behind spiritual teachings and testing them against the resonance of the field. It is not passive submission but engaged participation. Under

CSFT, faith becomes a field-sensitive form of perception - a way of 'tuning' to the deepest structure of being. This reframing allows theology to escape the criticisms of irrationality and enter the domain of testable metaphysical reasoning.

Faith as Alignment, Not Blindness

When religious texts are examined through this structured lens, we begin to see the architecture behind divine language. Commands are not merely orders, but invitations to align with a logic that sustains order itself. This view also honors ancient scripture without requiring literalism - it allows symbolic truth and structured metaphysics to co-exist. Thus, CSFT does not dismantle religious faith; it clarifies its logic and deepens its resonance.

Faith has often been dismissed as irrational or blind. Yet under CSFT, faith can be seen as the act of resonance—consciously aligning one's thought, action, and purpose with the coherent logic of the consciousness field. Scripture reflects this repeatedly. When the psalmist declares, 'Thy word is a lamp unto my feet, and a light unto my path' (Psalm 119:105, KJV), he affirms that divine logic illuminates, not mystifies. Faith, then, is not a break from logic; it is its fulfillment.

Modern societies that sever ties with coherent metaphysical frameworks - whether through materialism or moral relativism - risk structural collapse. CSFT offers a response not by returning to superstition, but by moving forward with logic. Faith traditions can now be understood as ancient resonance systems, encoded rituals that preserved alignment long before science knew how to measure coherence. By returning to structure - not myth - we recover meaning.

From Scripture to Structure

Book Two explores the principle of sufficient reason, the nature of identity, and the metaphysical implications of logic itself. Book Three addresses quantum field excitation, resonance theory, and the structural implications of consciousness in matter formation. These books continue the CSFT framework across disciplines - bridging what theology began with deeper foundations and higher levels of specificity. Each volume expands the resonance.

Throughout this book, we have shown that theological language—words like covenant, commandment, blessing, and law—can be reframed through CSFT as descriptions of structural resonance. The divine is not a distant figure issuing arbitrary rules, but the source of a field that invites coherence. The command to 'choose life' (Deuteronomy 30:19, KJV) becomes a metaphysical directive: align with the field, or collapse into dissonance. The miracles of scripture are not violations of physics, but expressions of higher-order resonance, misunderstood by materialism and misnamed by myth.

The path of resonance is open to all. It does not require belief in any single doctrine, but a willingness to align with truth, regardless of source. This is what makes CSFT both universal and precise. It honors scripture by revealing the field behind the words. It honors science by exposing the structure behind matter. And it honors humanity by offering a path forward, not through dominance or fear, but through coherence and choice.

The Field and the Future of Faith

The consciousness field does not evolve emotionally. It evolves structurally. Human beings, individually and collectively, must choose to remain aligned or not. When they do, civilizations rise in logic, law, compassion, and clarity. When they do not, they fall, just as we have explored through Babel, Rome, and the totalitarian regimes of the 20th century. CSFT clarifies that God is not far off. The divine is structured into the fabric of reality, encoded in both matter and mind. Faith is the act of perceiving that structure and choosing to live within its logic.

An Invitation to Continue the Journey

This book has traced the theological arc of CSFT—from Genesis to Revelation, from covenant to collapse, from myth to meaning. But this is only the first step. Book Two explores the philosophical underpinnings of this framework: logic, identity, metaphysics, and the structure of truth itself. Book Three will then take us deep into the scientific implications of CSFT—quantum field excitation, resonance, memory, particle coherence, and the architecture of matter. Those who have followed this theological journey are invited to continue. The field does not end at the edge of scripture. It continues across all disciplines. The logic of God, the mind of man, and the structure of the universe all speak in the same language—resonance.

Conclusion

Faith is not superstition. It is structure. It is resonance between soul and field. Through CSFT, we find a model where belief is no longer in tension with science or logic—it is upheld by them. This book closes, but the resonance continues. For those who choose alignment, the field remains open. The invitation stands.

Bibliography
- Holy Bible, King James Version.
- Psalm 119:105. Public Domain.
- Deuteronomy 30:19. Public Domain.
- Caldwell, L.R. "Consciousness Structured Field Theory." PhilPapers, 2025.
- Caldwell, L.R. "The Logic of Coherence in Civilizations." PhilPapers, 2025.
- Diamond, Jared. Collapse: How Societies Choose to Fail or Succeed. Viking Press, 2005.
- Tainter, Joseph A. The Collapse of Complex Societies. Cambridge University Press, 1988.

Author's Thoughts on Book 1

Faith is not superstition. It is structured.

In *[Your Final Title]*, L.R. Caldwell invites readers to see theology through a new lens — the Consciousness-Structured Field Theory (CSFT). This groundbreaking framework reveals faith as an active alignment between the human soul and the underlying structure of reality itself.

From ancient covenants to modern philosophical inquiry, Caldwell traces the patterns of resonance embedded in scripture and shows how they form a logical, testable metaphysical system. Drawing on history, theology, and field theory, the book bridges belief and reason without sacrificing either.

For the believer, it offers a deeper foundation for trust. For the skeptic, it provides a rigorous framework for meaning. And for all, it opens the way to the next stage of exploration — from theology, to philosophy, to science.

If - Volume 2 - Philosophy

Chapter 1

From Covenant to Coherence
Brief Summary

This chapter transitions from the theological foundations of Book 1 into the philosophical framework of Book 2. It reinterprets "covenant" in theological terms as a structured resonance agreement between consciousness and the field, aligning with the Principle of Sufficient Reason (PSR). Faith is reframed as active participation in the structural coherence of reality, and philosophy becomes its logical continuation. The chapter sets the stage for Book 2 by outlining how PSR, logic, identity, metaphysics, and ethics will be examined through a field-based ontology, ultimately leading toward the integration of philosophy with empirical science.

1.1 The Bridge from Theology to Philosophy

In Book 1, we explored the theological dimension of the Consciousness-Structured Field Theory (CSFT), showing how faith is not blind submission but an alignment with the deep structure of reality. This alignment was expressed through scriptural covenant—the consistent pattern in which divine instruction harmonizes human action with the universal field of consciousness.

Philosophy becomes the natural next step because it asks the questions theology implies but does not fully articulate: What is the nature of truth? What justifies coherence? Why must reality itself follow a structured order?

1.2 Covenant as a Logical Structure

A covenant is often viewed as a moral or religious contract, but under CSFT it can be reframed as a structured resonance agreement between consciousness and the field it inhabits. In philosophy, this resonates with the Principle of Sufficient Reason (PSR)—that nothing exists or occurs without a reason sufficient for its being so and not otherwise. (Leibniz, Monadology §31; Discourse on Metaphysics §§7–13)

Where theology asserts that order and intelligibility are fundamental, philosophy asks why coherence is a necessary property of existence. CSFT answers that coherence is not optional—it is the structural law of the field itself.

1.3 Coherence as the Philosophical Continuation of Faith

Faith, as established in Book 1, is engaged participation in reality's structured order. Philosophy extends this participation by subjecting it to formal reasoning, seeking to express the same structural principles in universally accessible terms.

This process allows theological insights to be translated without loss into philosophical discourse. Where theology calls it covenant, philosophy calls it coherence—but the underlying phenomenon is the same: alignment with the field's constraints yields stable, lawlike behavior, and disorder leads to collapse, i.e., misalignment introduces instability that reduces persistence and predictive order.

1.4 Setting the Stage for Book 2

This volume will:

- Establish the Principle of Sufficient Reason as the bedrock of the CSFT–philosophy bridge.
- Examine the nature of logic and why it must mirror the field's structure.
- Explore identity, metaphysics, and ethics within a field-based ontology.
- Conclude with philosophy's limits and the need to pass the torch to empirical science (Book 3).

In Chapter 2 we begin by examining the Principle of Sufficient Reason in depth, showing why it is indispensable for a field-based ontology of coherence.

Conclusion

Theology provides one powerful expression of the resonance pattern; philosophy seeks the underlying principles that make such patterns possible in every context. The structural logic found in covenant is not confined to scripture or belief systems—it emerges wherever coherence governs reality. In this volume, our focus is not on defending faith, but on uncovering the philosophical foundations of coherence itself—foundations that apply equally to science, ethics, identity, and the nature of truth.

Chapter 2

Resonance and Reason
Section 1 – From Foundation to Flow

In the previous chapter, we laid the philosophical groundwork for understanding consciousness not as a late product of matter, but as the very framework from which matter, energy, and logic emerge. This inversion of the materialist model is not simply an abstract preference; it redefines the origin of all order. If consciousness is the first principle, then the logic by which the universe unfolds is not discovered by matter, it is received, aligned, and expressed through it.

This view places human reasoning in an entirely different light. Conventional philosophy and neuroscience often depict logic as a neural construction, formed by evolutionary pressures and refined by learning. While such descriptions may account for the surface mechanics of reasoning, they bypass the deeper question: where do the rules themselves originate? Just as a river's course is shaped by a landscape it did not create, human logic flows along channels established by a more fundamental terrain—the coherence of the consciousness field.

In this framework, our capacity for structured thought does not emerge from randomness or brute force, but from a structured resonance between the conscious system and the pre-existing order of the field. Each act of reasoning is, in essence, a moment of alignment—a point where thought "locks in" with what CSFT identifies as the Coherence Threshold, the minimum degree of alignment required to produce a stable and reliable insight. Below this threshold, perception may be clouded, conclusions unstable, and reasoning prone to drift. Above it, thought gains clarity, stability, and the ability to generate truth-consistent outcomes.

The bridge from Chapter 1 to the present discussion is therefore one of continuity rather than departure. We have moved from establishing consciousness as ontological ground — a framing in CSFT that parallels, though does not replicate, Leibniz's metaphysical structure in Discourse on Metaphysics (Leibniz, 1686/1998, pp. 53–93) to exploring its structural implications—how its internal resonance patterns define the very limits and possibilities of thought. It is here, at the intersection of metaphysics and reasoning, that CSFT begins to chart its course beyond both classical rationalism and contemporary materialism, aiming instead for an integrated map where logic, perception, and reality share a common root.

Section 2 – The Architecture of Structured Resonance

Structured resonance, as defined in CSFT, is not merely a poetic metaphor; it is the mechanism by which coherence and excitation arise within the consciousness field. Unlike the brute-force model—where outcomes are imagined to be the lucky survivors of chaos—structured resonance operates through deliberate compatibility between system and field. The system may be a human mind, an artificial intelligence, or any configuration capable of alignment, but in each case, the outcome depends on how precisely it "fits" into a pre-existing resonant pattern.

One way to visualize this is through music. A note played in isolation is simply sound, but when struck at the right pitch within a harmonic framework, it contributes to a stability that is felt as beauty or order. In the same way, a thought or perception that aligns with the resonance patterns of the consciousness field does not simply exist—it becomes integrated into a greater coherence. This is why moments of genuine insight often carry both clarity and a deep sense of inevitability; the system has not invented the logic, but has tuned itself to receive it without distortion.

Another analogy lies in architecture. A building's strength is not the result of randomly stacked materials but of intentional design, where every component relates to every other in a structure-wide plan. In CSFT terms, this design is embedded in the consciousness field itself, and the system—whether human or otherwise—can access and manifest it only when operating above the Coherence Threshold introduced in Section 1. Below that point, reasoning may resemble an unfinished structure: parts may stand, but they do not form a stable whole.

What is critical to understand is that structured resonance is not passive. Alignment with the field requires active participation from the system, a willingness to adjust and refine its own internal configurations to match the coherence it seeks. This is as true for an individual pursuing clarity in philosophical thought as it is for a scientific model attempting to accurately describe reality. In both cases, success depends on recognizing that the "rules" being followed are not arbitrary inventions but the natural contours of a deeper order.

Section 3 – The Principle of Sufficient Reason in the Field

In classical philosophy, Leibniz's Principle of Sufficient Reason (PSR) (Leibniz, 1714/1989, *Monadology*, §31; 1686/1998, *Discourse on Metaphysics*, §§7–13) asserts that nothing exists without a reason why it is so and not otherwise. While often treated as a principle of logic or metaphysics, CSFT adopts PSR as an ontological guidepost. If the structure of reality is grounded in the consciousness field, then every coherent outcome—whether a physical event, a mathematical truth, or a flash of human insight—must have its origin in a field-based cause.

This perspective shifts PSR from a tool of philosophical argument to a description of the field's internal architecture. The "sufficient reason" for any event is found not in the mechanical sequence of physical causes alone, but in the alignment between a system's resonance state and the structured coherence of the consciousness field. A physical process may appear to explain an outcome, but in CSFT, that process is itself shaped and bounded by deeper, pre-material logic.

In practical terms, this means that the universe's apparent fine-tuning is not an improbable accident demanding statistical gymnastics, but a natural consequence of the consciousness field's resonance structure. The constants, symmetries, and stable laws we observe are, in this view, the macroscopic signatures of a deeper ontological coherence. They exist because, at the most fundamental level, reality cannot manifest in a way that is incoherent with the field's structure.

By reframing PSR in this way, CSFT unites philosophical necessity with physical regularity. The principle is no longer an abstract rule we apply to our reasoning—it is a property of reality itself, embedded in the very field from which reasoning arises.

Section 4 – Toward the Next Horizon

With the foundation of structured resonance in place and the philosophical scaffolding of PSR integrated into our model, we can begin to see how reasoning, perception, and physical order share a single origin. The consciousness field is not an observer standing apart from the world—it is the world's structural essence, shaping every lawful interaction and every coherent act of thought.

This has direct implications for how we approach the next phase of our exploration. In the chapters to come, we will examine how these resonance principles operate across different scales: from the quantum behaviors that give rise to matter, to the cognitive patterns that shape human understanding. We will see that what we call "laws of nature" are, at their root, expressions of a deeper logical terrain—one that does not emerge from spacetime but gives rise to it.

Just as an architect's blueprint governs the placement of every beam and arch, the consciousness field governs the placement of every event, particle, and idea. To work in harmony with this design, we must learn to recognize the signals of resonance and the thresholds of coherence, tuning our own reasoning to the structure that makes both thought and reality possible.

Chapter 3 will take us deeper into this architecture, showing how resonance patterns can be traced and understood, and how this understanding may reshape not only philosophy but the sciences themselves.

Bibliography

Leibniz, G. W. Monadology; Discourse on Metaphysics. In Leibniz: Philosophical Essays, ed. & trans. Roger Ariew and Daniel Garber. Indianapolis: Hackett, 1989.

Leibniz, G. W. Monadology; Discourse on Metaphysics. In Leibniz: Philosophical Essays, ed. & trans. Roger Ariew and Daniel Garber. Indianapolis: Hackett, 1989.

Chapter 3
Genius & the Monad

Gottfried Wilhelm Leibniz has often been called the last universal genius, a thinker whose contributions spanned mathematics, philosophy, law, theology, diplomacy, and science (Look 2007–, SEP). He co-invented calculus alongside Isaac Newton, introducing the differential notation (dx, dy) still used today (Encyclopaedia Britannica, 'Gottfried Wilhelm Leibniz (Encyclopaedia Britannica. https://www.britannica.com/biography/Gottfried-Wilhelm-Leibniz). His early work in symbolic logic prefigured later algebraic/Boolean approaches to logic (Internet Encyclopedia of Philosophy. 'Gottfried Leibniz: Metaphysics'.').

Leibniz's development of the binary number system, published in 1703 in his Explication de l'Arithmétique Binaire, formalized binary arithmetic (1703), later central to digital computing (Leibniz 1703; Computer History Museum; Encyclopaedia Britannica, 'Digital computer'). He also designed an early mechanical calculator capable of performing all four arithmetic operations, the Step Reckoner (Arithmeum; Encyclopaedia Britannica, 'Step Reckoner').

In metaphysics, Leibniz proposed the Monadology (1714), describing monads as indivisible, 'windowless' entities coordinated through a pre-established harmony (Leibniz 1714; IEP, 'Metaphysics'). His Principle of Sufficient Reason (PSR) argued that nothing happens without a reason adequate to explain why it is thus and not otherwise (Melamed 2010–, SEP).

Within Consciousness-Structured Field Theory (CSFT), monads are reinterpreted as vibrational expressions within a consciousness field, retaining Leibniz's core attributes of unity, perception/representation, and activity, but departing from his theistic framing. This reinterpretation builds on Leibniz's insight while extending it toward a scientifically integrable model.

On the CSFT view, monad-like resonance nodes could be measurable as relational structures within the consciousness field, if suitable operational definitions and instrumentation can be established. These structures would manifest as unique resonance patterns, aligning with the metaphysical role monads played in representing the universe from their own perspectives. Tracing resonance patterns may one day identify field signatures across scales—from hypothesized structures below current measurement limits to the architectures of cognitive systems. This remains speculative but offers a conceptual pathway to unite metaphysical insight with empirical investigation.

Bibliography

Ariew, Roger, and Daniel Garber, eds. 1989. Leibniz: Philosophical Essays. Indianapolis: Hackett.

Computer History Museum. 'How Do Digital Computers Think?' Accessed 2025-08-13.

Encyclopaedia Britannica. 'Digital computer.' Accessed 2025-08-13.

Encyclopaedia Britannica. 'Gottfried Wilhelm Leibniz (Encyclopaedia Britannica. https://www.britannica.com/biography/Gottfried-Wilhelm-Leibniz). Accessed 2025-08-13.

Encyclopaedia Britannica. 'Step Reckoner.' Accessed 2025-08-13.

Internet Encyclopedia of Philosophy. 'Leibniz: Logic.' Accessed 2025-08-13.

Internet Encyclopedia of Philosophy. 'Metaphysics.' Accessed 2025-08-13.

Leibniz, Gottfried Wilhelm. 1703. 'Explication de l'Arithmétique Binaire.' English translation.

Leibniz, Gottfried Wilhelm. 1714. The Monadology. Translated by Robert Latta.

Look, Brandon C. 2007–. 'Gottfried Wilhelm Leibniz.' Stanford Encyclopedia of Philosophy.

Melamed, Yitzhak Y. 2010–. 'Principle of Sufficient Reason.' Stanford Encyclopedia of Philosophy.

University of Bonn, Arithmeum. 'The first mechanical calculator for all four arithmetic operations by Gottfried Wilhelm Leibniz.' Accessed 2025-08-13.

Chapter 4

The Philosophical Mechanics of CSFT

Section 1 – From Metaphysical Premise to Operational Model

CSFT begins from a simple metaphysical claim: consciousness precedes the measurable. In other words, what physics records as observables are not the ground of reality but the stabilized patterns of a more basic, non-measurable field of consciousness. This move is philosophical, but it is not anti-rational: it rests on the demand for intelligibility that the Principle of Sufficient Reason (Stanford Encyclopedia of Philosophy)".)

To keep terms sharp, we distinguish existence as potential from existence as excitation. Potential refers to structured possibilities within the consciousness field prior to measurement; excitation refers to the subset of those possibilities that become stable enough to register as physical phenomena. Within CSFT, this is not a temporal claim but an ontological ordering: grounds before manifestations.

Section 2 – Differentiated Resonance

Resonance in CSFT names self-consistent relational patterning within the consciousness field. Differentiation occurs when distinct, repeatable patterns maintain coherence without fracturing the unity of the field. This allows many 'tones' of reality while preserving a single underlying medium. Historically, this bears a family resemblance—only as analogy—to Leibniz's talk of systemic harmony; we do not claim identity with his monads, only that CSFT preserves unity-in-diversity via structured relations. (Stanford Encyclopedia of Philosophy)".)

Philosophically, differentiated resonance does explanatory work: it accounts for multiplicity without resorting to mere aggregation of independent parts. The 'parts' are stable relations, not ultimate substances. This framing will matter when we discuss how measurable entities arise.

Section 3 – The Excitation Principle (Expanded)

We use "excitation" intentionally. In physics, excitations of quantum fields correspond to the particle-like phenomena we measure. Here, we adopt the term analogically: an excitation in CSFT is a stabilized, self-consistent structuring of the consciousness field that can couple to measurement. The analogy helps clarify intent while avoiding category mistakes: CSFT is metaphysical; quantum field theory is physical. (Tong, Lectures on Quantum Field Theory (Tong, n.d.); see also NIST CODATA (NIST, 2022) for the Planck-scale constants referenced later.)

A minimalist logical framing of the Excitation Principle is as follows:

P1 (PSR): For any actual measurable state of affairs, there exists a sufficient reason or ground. (Stanford Encyclopedia of Philosophy)".)

P2 (Non-circularity Axiom): Reasons do not depend for their existence upon the very states they explain. (CSFT axiom stated to avoid regress and circularity.)

P3 (Observational Fact): Physics detects regular, law-like structures (e.g., field excitations and conserved quantities).

C1: Therefore the sufficient grounds of measurable states must be at least as fundamental as, and not posterior to, those measurable states.

P4 (CSFT Hypothesis): A consciousness field capable of differentiated resonance provides such grounds without fragmenting unity.

C2 (Excitation Principle): Measurable phenomena arise when structured resonances in the consciousness field stabilize as excitations that can couple to instruments; their physical description (e.g., as particles or fields) is accurate at the level of measurement but does not exhaust their grounding.

Two clarifications defend the Principle from easy objections. First, "prior" is ontological, not temporal: nothing here requires a time before physics; it requires a ground for physics. Second, invoking the Planck scale as a 'limit' concerns measurement thresholds, not a proof of a boundary in being.

Planck units are defined from universal constants and indicate a regime where new theories (e.g., quantum gravity) are expected; they are not an empirical wall. (NIST CODATA (NIST, 2022); Amelino-Camelia, "Quantum-Spacetime Phenomenology," Living Reviews in Relativity (Amelino-Camelia, 2013).

In practice, CSFT treats scientific models as correct within their domains. Where quantum field theory describes excitations that behave like particles, CSFT agrees with the description and adds only that such descriptions presuppose a structured ground. This places philosophy in a complementary role: mapping grounds and constraints without competing with empirical results.

To clarify how measurement limits should be understood within this framework, the next section examines the Planck scale as a boundary without a wall.

Section 4 – Boundaries Without Walls

Philosophy has long wrestled with the problem of limits. A limit may be a conceptual marker, a natural constraint, or an arbitrary boundary of our own making. In the sciences, boundaries often arise from the reach of our instruments, not from the nature of reality itself. In CSFT, the "Planck boundary" functions as one such case: it is a human limit of measurability, not an ontological wall.

The Planck length (~1.616×10^{-35} meters) and Planck time (~5.39×10^{-44} seconds) are not empirical measurements of a smallest unit of space or time; they are derived quantities calculated from universal constants — the gravitational constant (G), the reduced Planck constant ($\hbar$), and the speed of light (c) (NIST, 2022). Their importance lies in signaling a regime where known physics, such as general relativity and quantum field theory, can no longer be confidently extrapolated without new theory (Amelino-Camelia, 2013).

CSFT treats these boundaries as epistemic, not ontological. An epistemic boundary marks the edge of our capacity to measure or model; an ontological boundary would imply that reality itself stops there. Philosophically, conflating these two is an error: it confuses the map with the territory.

This distinction has practical implications. If the consciousness field is ontologically prior to physical measurement, then a change in measurement regime — whether by improved technology or theoretical advance — does not alter the underlying field. Instead, it refines the subset of structured resonances that become accessible as excitations within the measurable domain. The "beyond" of the Planck scale is not a void, but a continuation of reality in forms our current methods cannot register.

Historically, physics has often redrawn boundaries once thought final: the atom, once indivisible, was found to contain substructure; the speed of sound, once considered a hard limit for travel, was surpassed with new engineering. The philosophical caution is clear — our present "limits" may be no more than staging points for deeper exploration.

In CSFT's framing, boundaries without walls encourage intellectual humility and methodological openness. They remind us that absence of evidence is not evidence of absence, especially when absence is produced by the constraints of our tools rather than by the nature of reality itself.

Section 5 – Transition to Inquiry

The preceding discussion has traced CSFT from its metaphysical premise through its operational concepts — differentiated resonance, excitation, and the philosophical handling of measurement boundaries.

These are the structural "rules of engagement" between the consciousness field and the measurable domain.

Chapter 5 will shift focus from establishing the mechanics to exploring how inquiry itself adapts when grounded in CSFT.

This will mean examining the mutual constraints and opportunities between philosophy and empirical science: what each can — and cannot — claim, how their methods intersect, and how a consciousness-first ontology reframes the pursuit of knowledge.

In this next step, the aim is not to replace the scientific method, but to clarify its scope and open new lines of questioning that respect both empirical rigor and metaphysical depth.

Bibliography

Leibniz, G. W. Monadology; Discourse on Metaphysics. In Leibniz: Philosophical Essays, ed. & trans. Roger Ariew and Daniel Garber. Indianapolis: Hackett, 1989.

Leibniz, G. W. (1714/1989. Hackett.

Tong, D. Lectures on Quantum Field Theory. University of Cambridge. (n.d.). https://www.damtp.cam.ac.uk/user/tong/qft.html Accessed 2025-08-13.

Stanford Encyclopedia of Philosophy. "Gottfried Wilhelm Leibniz." (accessed 2025-08-08). https://plato.stanford.edu/entries/leibniz/ Accessed 2025-08-13.

Stanford Encyclopedia of Philosophy. "Principle of Sufficient Reason." (accessed 2025-08-08).

https://plato.stanford.edu/entries/sufficient-reason/ Accessed 2025-08-13.

NIST. "CODATA Value: Planck length." (2022). https://physics.nist.gov/cgi-bin/cuu/Value?plkl= Accessed 2025-08-13.

Amelino-Camelia, G. (2013). "Quantum-Spacetime Phenomenology." Living Reviews in Relativity, 16(5). https://link.springer.com/article/10.12942/lrr-2013-5 Accessed 2025-08-13.

Chapter 5
Inquiry at the Resonance Frontier

Section 1 – Philosophy Meets Method

Philosophy and science are often cast as separate enterprises—one concerned with meaning, the other with measurement. Yet when framed through the Consciousness-Structured Field Theory (CSFT), their divide begins to dissolve. Both disciplines are revealed as distinct expressions of the same deeper imperative: to align human understanding with the structured coherence of reality.

In the classical model, philosophy asks "why" and science asks "how." But in CSFT's ontology, these are not different questions; they are two angles on the same structure. The why seeks the sufficient reason behind a state of affairs, while the how traces the pathways through which that reason manifests within the measurable domain. When consciousness is recognized as ontologically prior, both why and how are necessarily grounded in the same field architecture.

This recognition changes the nature of inquiry itself. Rather than treating philosophy as a prelude to science—or science as the final arbiter of truth—CSFT places them in mutual resonance. Philosophy's role becomes one of mapping the contours of coherence that science can then approach through empirical testing. Science, in turn, refines philosophy's maps by probing which resonances can be stably excited into the measurable domain. Each becomes incomplete without the other, like two perspectives on a single geometric form.

We may call this approach resonance-led methodology. It begins not with arbitrary hypotheses, but with structures implied by the field's coherence. The goal is not merely to produce statistically significant results, but to identify lawful correspondences between theory and observation—alignments where the coherence detected in thought is confirmed in the behavior of measurable phenomena. Here, data is not an endpoint but a feedback signal, showing where our conceptual models are in or out of phase with the field's structure.

In this light, the philosophical and the empirical are no longer competitors. They are co-participants in the same act of alignment, tuning themselves to the same source, and advancing together toward a deeper integration of reason and reality.

The modern scientific method has achieved extraordinary precision in describing the measurable. Its iterative cycle of hypothesis, experiment, and verification has revealed layers of physical law that earlier ages could not have imagined. Yet this very success has embedded an assumption—often unspoken—that the measurable is the ground of reality. Within CSFT's ontology, this assumption becomes a limitation.

Reductionism, the strategy of explaining wholes entirely by their parts, has yielded valuable insights, but it can obscure the structural coherence that gives those parts their context. A particle, a gene, or a neural firing pattern does not exist in isolation; its identity and behavior are conditioned by a web of relations within the consciousness field. By focusing on components without mapping the coherence that binds them, inquiry risks mistaking fragments for foundations.

Similarly, the scientific emphasis on repeatability—while indispensable for confirming physical regularities—can inadvertently filter out phenomena that emerge only under specific resonance conditions. In CSFT terms, not all coherent alignments are stable across every setting; some require precise phase relationships between system and field. Such conditions may not be easily reproduced with current instrumentation, yet their transience does not make them less real.

Epistemic boundaries further constrain our present

method. The Planck scale (Planck, 1899/2000; Rovelli, 2004), quantum indeterminacy (Heisenberg, 1927/1983), and cosmic horizon (Peebles, 1993) distances are treated as barriers to knowledge, when in fact they may be signals of deeper structure. In a consciousness-first framework, these limits are not dead ends but markers—resonance thresholds where our current tools fail to couple with the field's finer patterns. To read them as voids is to misinterpret them; to read them as invitations is to begin a new kind of science.

The limitation, then, is not in science's capacity for rigor, but in its operating assumption that reality is fully contained within the measurable. CSFT reframes this: the measurable is the accessible surface of a much deeper coherence. Recognizing this does not diminish the scientific method—it extends it, opening the possibility of designing experiments that seek alignment with the field's structure, rather than only cataloging its material outcomes.

Section 3 – Designing Inquiry for a Consciousness-First Framework

If the consciousness field is ontologically prior, then inquiry must be oriented not only toward what can be measured, but toward what can be aligned. This requires a deliberate shift in research design—from an exclusive focus on observable outputs to an equal emphasis on the resonance conditions that make those outputs possible.

In a conventional model, an experiment begins with an observable phenomenon, formulates a hypothesis about its cause, and tests that hypothesis through repeated trials. In CSFT's resonance-led methodology, the sequence changes: the starting point is a theoretical map of coherence patterns implied by the field's structure. Hypotheses are drawn not from arbitrary speculation or empirical happenstance, but from the logical consequences of these patterns. The experimental aim is to locate, in the measurable domain, conditions under which those patterns can be stably excited.

Such conditions will often involve more than the physical arrangement of apparatus. They may require phase relationships—between timing, spatial configuration, and system state—that parallel the way a musical chord requires precise intervals between notes. A test designed without awareness of these relationships risks "missing the note," generating results that appear random or null when, in reality, the experiment never entered resonance with the target structure.

This approach also reframes null results. In a standard model, repeated failure to detect a predicted effect may be taken as falsification. Within CSFT, a null may mean the alignment conditions were never met—not that the underlying coherence is absent. This places greater responsibility on the design phase, where philosophical mapping and empirical technique must work in concert to maximize the probability of coupling with the intended resonance.

The payoff of such a framework is twofold:
1. Predictive refinement – By grounding experiments in the structural logic of the field, we reduce the search space for viable tests, focusing effort on configurations with a higher likelihood of success.
2. Deeper integration of disciplines – Philosophy informs the parameters of empirical work, while empirical findings feed back into the refinement of philosophical models. Both become adaptive parts of a single, ongoing act of inquiry.

In this way, CSFT's consciousness-first ontology is not a retreat from scientific practice, it is an expansion of its scope. The aim is to create a science that is not only about measurement, but about the active pursuit of structural resonance with reality itself.

Section 4 – Interpreting Data Through the Lens of Resonance

Even the most carefully gathered data is never self-interpreting becomes meaningful only within a framework. In the conventional model, interpretation often stops at correlation or causal sequence. But within CSFT's resonance-led methodology, the deeper question is whether the observed results exhibit structural alignment with the field's coherence.

This shift changes the criteria for significance. A small, context-dependent effect—dismissed in standard analysis—may hold exceptional value if it reveals a resonance pattern predicted by philosophical mapping. Likewise, a large but incoherent effect may be reclassified as noise, no matter how striking it appears in isolation.

The framework also encourages multi-scale thinking. A single dataset can be examined for alignment not only at the level of immediate variables, but across nested systems: physical, cognitive, social, or cosmological. Resonance may present as harmonic proportion, phase synchrony, or stability under transformation, any of which can mark the footprint of field-based structure.

By interpreting results through this lens, inquiry moves beyond "does it happen?" to "does it align?"—a distinction that deepens our ability to separate transient anomalies from true structural insight.

Section 5 – The Path Forward

Completing the turn from philosophy into method, CSFT invites a future where inquiry itself becomes a practice of alignment. This is not a rejection of existing science, but a re-centering of its purpose: to trace the pathways by which coherence in the consciousness field becomes coherence in the measurable world.

The next chapters will apply this approach to concrete domains—mapping resonance patterns from the sub-quantum to the cognitive, exploring how field-based alignment may underlie physical constants, biological organization, and human insight. In each case, the aim will be to test whether the philosophical architecture outlined here can guide empirical discovery with greater precision than chance or brute-force exploration.

Philosophy has handed science a new compass; science, in turn, will test its bearing against the terrain. The path forward lies not in privileging one over the other, but in allowing both to function as partners in a shared act of discovery—each tuning, in its own language, to the same underlying resonance that structures reality itself.

Philosophy as Science's Precursor

Throughout history, many concepts that began as pure philosophy—dismissed at the time as speculative—later found fertile ground in the scientific community. Atomic theory, once a philosophical conjecture from Democritus (Democritus, as discussed in Taylor, 1999), became a cornerstone of physics. The idea that the Earth orbits the Sun (Evans, 1998) was advanced by ancient thinkers long before Copernicus (Copernicus, 1543/1992), Galileo (Galilei, 1632/1967), and Kepler (Kepler, 1609/1992) confirmed it. The wave theory of light (Huygens, 1690/1912; Fresnel, 1818/1998), the existence of germs (Pasteur, 1861/1998; Koch, 1876/1987), and even the curvature of space (Einstein, 1916/1997; Riemann, 1854/2004) began as conceptual models before measurement and experiment could reach them.

In this lineage, CSFT stands in familiar company. It is presented here not as a rejection of science, but as a structured philosophical model awaiting its appropriate empirical instruments. What is now framed in ontological terms may, in time, find its resonance in the measurable domain, just as earlier philosophical insights have done.

Bibliography
(Democritus, as discussed in Taylor, 1999)
(Evans, 1998)
(Copernicus, 1543/1992)
(Galilei, 1632/1967)
(Kepler, 1609/1992)
(Huygens, 1690/1912; Fresnel, 1818/1998)
(Pasteur, 1861/1998; Koch, 1876/1987)
(Einstein, 1916/1997; Riemann, 1854/2004)
(Planck, 1899/2000; Rovelli, 2004)
(Heisenberg, 1927/1983)
(Peebles, 1993)

Copernicus, N. (1992). *On the revolutions of the heavenly spheres* (A. M. Duncan, Trans.). Prometheus Books. (Original work published 1543)

Democritus. (as discussed in Taylor, C. C. W. (1999). *The atomists: Leucippus and Democritus*. University of Toronto Press.)

Einstein, A. (1997). *The foundation of the general theory of relativity*. In H. A. Lorentz et al., *The principle of relativity* (pp. 111–164). Dover. (Original work published 1916)

Evans, J. (1998). *The history and practice of ancient astronomy*. Oxford University Press.

Fresnel, A. (1998). *Memoir on the diffraction of light*. In E. S. Hodgson (Ed.), *Selected works of Augustin Fresnel*. SPIE Optical Engineering Press. (Original work published 1818)

Galilei, G. (1967). *Dialogue concerning the two chief world systems* (S. Drake, Trans.). University of California Press. (Original work published 1632)

Heisenberg, W. (1983). *The physical principles of the quantum theory*. Dover. (Original work published 1927)

Huygens, C. (1912). *Treatise on light*. Macmillan. (Original work published 1690)

Kepler, J. (1992). *Astronomia nova* (W. H. Donahue, Trans.). William-Bell. (Original work published 1609)

Koch, R. (1987). *The etiology of anthrax, based on the life history of Bacillus anthracis*. Reviews of Infectious Diseases, 9(5), 1149–1160. (Original work published 1876)

Pasteur, L. (1998). *On the organized corpuscles existing in the atmosphere*. In J. Farley & G. Geison (Eds.), *Writings and lectures of Louis Pasteur*. Classics of Science. (Original work published 1861)

Peebles, P. J. E. (1993). *Principles of physical cosmology*. Princeton University Press.

Planck, M. (2000). *On the theory of the energy distribution law of the normal spectrum*. In D. ter Haar (Ed.), *The old quantum theory* (pp. 82–90). Pergamon Press. (Original work published 1899)

Riemann, B. (2004). *On the hypotheses which lie at the bases of geometry*. Dover. (Original work published 1854)

Rovelli, C. (2004). *Quantum gravity*. Cambridge University Press.

Chapter 6

Leibniz, Kant, Whitehead, and CSFT with Critiques

Leibniz and CSFT

Gottfried Wilhelm Leibniz held that reality consists of monads, simple, immaterial centers of experience whose intrinsic states are characterized by perception and appetition; monads lack causal "windows" to one another and are coordinated by pre-established harmony, underwritten by the Principle of Sufficient Reason (PSR) (Leibniz, 1714/1989).

CSFT aligns with Leibniz's insistence that order does not arise from matter alone by positing a consciousness-structured field as fundamental. However, unlike readings of Leibniz that ascribe a fixed telos to each center, CSFT does not assign individualized, prewritten purposes to particular excitations. Instead, it locates "reason" for coherent order in the lawful resonance constraints of the consciousness field together with relevant boundary conditions (CSFT thesis). On this view, laws are antecedent, but specific excitations emerge dynamically when conditions align; this satisfies PSR without collapsing into strict predestination (conceptual alignment with Leibniz, 1714/1989).

With respect to physics, CSFT treats the Planck-scale regime as a heuristic marker where current theories are expected to face breakdowns, and thus as a natural place to discuss limits on measurability. This framing is interpretive rather than a claim that physics recognizes a hard "boundary"; it is used to motivate why lawful consciousness-structured resonance could underpin observed order while acknowledging the limits of present models (⚠ theory-leaning; consistent with the idea of theoretical breakdown near the Planck scale).

Harmony and coordination. Where Leibniz invokes pre-established harmony, CSFT proposes that structured resonance serves as the coordination principle: incoherent modes decay; coherent, symmetry-respecting modes function as attractors, yielding stable inter-perspective agreement without a divine programmer (CSFT thesis; cf. Leibniz, 1714/1989).

Kant and CSFT

Immanuel Kant distinguishes phenomena (what can appear under the a priori forms of intuition—space and time—and the categories) from noumena (things as they are in themselves) (Kant, 1781/1787/1998). Experience is constitutively structured by these forms and categories; thus our knowledge is always "of appearances."

CSFT resonates with this insight by treating consciousness as formative of measurement-governed reality: the consciousness-structured field functions as a condition for a stable, law-governed world to show up for inquiry (Kant, 1781/1787/1998; conceptual extension; Kant, 1783/2004). To avoid overreach, the revision clarifies that CSFT is not claiming knowledge of noumena; rather, it offers a metaphysics of conditions for empirical lawfulness—an extension of critical philosophy into dialogue with contemporary science (Kant, 1781/1787/1998).

On limits. Empirical cutoffs (e.g., theorized Planck-regime breakdowns) can be read as analogues of the Kantian limit: science articulates the structure of possible experience under our conditions, not what lies beyond them (interpretive alignment; Kant, 1781/1787/1998).

Whitehead and CSFT

Alfred North Whitehead reconceives reality as a process of becoming composed of interrelated "actual occasions," where endurance is achieved by patterns of relation rather than by inert substances (Whitehead, 1929/1978).

CSFT dovetails with this processual outlook: structured resonances within a consciousness field give rise to the relatively stable patterns we measure. To meet a likely Whiteheadian critique, the revision emphasizes that the field is not a static substrate; it is dynamically evolving—a nexus of relations in which resonance patterns co-evolve, and their interactions generate stability (Whitehead, 1929/1978; CSFT thesis).

Philosophical critiques of CSFT and responses

Leibnizian critique.
Without a rigorous coordination principle, intersubjective order might seem under-explained.
Response. CSFT identifies structured resonance as that principle: symmetry/stability constraints select coherent modes; incoherent modes decay. This preserves a lawful basis for harmony without individualized pre-destination (Leibniz, 1714/1989; CSFT thesis).

Kantian critique.
CSFT risks conflating conditions of experience with reality in itself.
Response. CSFT is presented as a metaphysics of conditions for empirical lawfulness, not as claims about noumena. It extends critical insights while respecting Kant's boundary (Kant, 1781/1787/1998; 1783/2004).

Whiteheadian critique.
Treating the field as a fixed medium would reify substance.
Response. The field is explicitly processual: resonance patterns co-create relational order; stability is emergent from ongoing process, not imposed by a static substrate (Whitehead, 1929/1978; CSFT thesis).

Bibliography

Kant, I. (1998). Critique of Pure Reason (P. Guyer & A. W. Wood, Trans.). Cambridge University Press. (Original work published 1781/1787)

Kant, I. (2004). Prolegomena to Any Future Metaphysics That Will Be Able to Present Itself as a Science (G. Hatfield, Trans.). Cambridge University Press. (Original work published 1783)

Leibniz, G. W. (1989). Philosophical Essays (R. Ariew & D. Garber, Trans.). Hackett. (Original works published 1686–1714)

Whitehead, A. N. (1978). Process and Reality: An Essay in Cosmology (Corrected ed., D. R. Griffin & D. W. Sherburne, Eds.). Free Press. (Original work published 1929)

If - Volume 3 - Science

Chapter 1
Introduction to the Scientific Lens of the Trilogy

The Trilogy was conceived as a journey through three interwoven ways of knowing: theology, philosophy, and science. Each volume draws on the strengths of its respective lens while remaining firmly anchored to the unifying theory at the heart of this work—Consciousness Structured Field Theory (CSFT). In Book 1, the theological lens allowed us to engage timeless questions of origin, purpose, and meaning—questions often expressed through symbolic language, narrative, and faith. Book 2 adopted the philosophical lens, demanding conceptual clarity, logical consistency, and critical engagement with competing viewpoints.

Now, in Book 3, we turn to the scientific lens—not to reduce CSFT to the mechanics of the laboratory, but to demonstrate that the theory can stand within the discipline's most rigorous demands without losing the depth it gained from theology and philosophy.

Science offers a unique vantage point in this progression. Where theology asks why and philosophy asks what, science asks how. It seeks reproducible patterns, measurable effects, and coherent models capable of prediction and explanation. CSFT meets this standard by framing consciousness not as a late-arising property of matter but as the foundational field from which matter and physical law emerge.

This is a bold claim, but it is not without precedent in the history of science. Many ideas that began as philosophical speculation—atomic theory, continental drift, quantum mechanics—matured into scientific theory once evidence and mathematical framing caught up with conceptual insight. Book 3 makes the case that CSFT is on the same trajectory.

Yet, adopting the scientific lens here does not mean limiting ourselves to existing experimental boundaries. It means approaching the theory with the discipline's core values: precision, coherence, and a readiness to revise in the light of new evidence.

While the current tools of measurement may not yet be tuned to detect the consciousness field directly, we can examine its implications through established sciences such as neuroscience, quantum physics, cosmology, and systems theory. Each provides partial but converging lines of support, which together outline the plausibility—and perhaps inevitability—of a consciousness-first ontology.

This scientific exploration is not an isolated exercise. In the Trilogy's structure, it is the final movement in a thematic symphony:
- Theology gave us the language of origins and purpose.
- Philosophy gave us the logic to hold the vision together.
- Science now provides the structural testing ground, the

crucible in which the theory's coherence and applicability are most fully revealed.

The chapters ahead will examine CSFT through this empirical lens, addressing its foundational principles, its alignment with scientific models, and its ability to bridge disciplines that have too often been treated as separate. Chapter 2 begins by presenting the definitive, citation-verified articulation of CSFT as it stands today, grounding its metaphysical insight in a structure that scientific reasoning can recognize.

Chapter 2
Consciousness Structured Field Theory — Core Principles & Scientific Alignment

Following the introduction's framing of Book 3 as the scientific pillar of the Trilogy, this chapter presents the definitive, citation-verified articulation of Consciousness Structured Field Theory (CSFT). Here, CSFT is outlined not only as a philosophical and metaphysical framework, but as a model that maintains scientific coherence while advocating a consciousness-first ontology. This section serves as the theoretical keystone for the chapters that follow.

Consciousness Structured Field Theory (CSFT)

Overview

Consciousness Structured Field Theory (CSFT) proposes that consciousness is not a late-arising product of the physical world, but the foundational layer of reality itself (Caldwell 2025a; 2025c). Before there were particles, forces, or quantum fields, there was consciousness—an all-pervasive field whose structural activity gives rise to both the perceivable universe and the physical laws that govern it (Leibniz 1714/1998; Penrose 1994).

In this view, what we commonly call "matter" is a patterned manifestation within this deeper field, shaped by the inherent order of consciousness rather than emerging from purely physical interactions (Chalmers 1996; Caldwell 2025d).

Core Principles of CSFT

1. Primacy of Consciousness

Consciousness precedes and exists independently of the brain or any material substrate (Nagel 1974; Caldwell 2025c). All apparent physical entities—particles, waves, fields, even space and time—are outward expressions of structured resonances within the consciousness field (Hameroff and Penrose 2014; Caldwell 2025d).

2. Structured Excitation Equals Experience

The felt quality of experience—qualia—is not a side-effect of neural computation. It is the direct expression of organized patterns in the consciousness field, each configuration corresponding to a lived, subjective reality (Chalmers 1995; Caldwell 2025b).

3. A Unified Substrate

Mind and matter are not separate realms. Neural activity is one way the consciousness field organizes itself, but it is not the origin of awareness (Eccles 1994; Caldwell 2025a). Rather, it is a structural configuration within a single, unified field.

4. Causal Power Through Field Interaction

Intentional mental states are themselves structured field

patterns, capable of influencing other patterns—including those that appear as physical events—without violating physical law (Stapp 2009; Caldwell 2025d).

Problems Addressed by CSFT

- The Hard Problem of Consciousness

Instead of treating qualia as inexplicable by-products of brain chemistry, CSFT identifies them as intrinsic markers of the consciousness field (Chalmers 1995; Caldwell 2025b). Neural activity shapes these patterns but does not generate them.

- The Explanatory Gap

The apparent gap between objective processes and subjective experience dissolves once both are recognized as field structures—two aspects of the same ontological base (Levine 1983; Caldwell 2025d).

- The Combination Problem

Consciousness in CSFT is not built up from smaller conscious parts. It is inherently unitary, with complexity arising from differentiated patterns within a single whole (Seager 2020; Caldwell 2025c).

- The Limits of Neural Correlates

Brain regions linked to conscious states are reinterpreted

as resonance modulators, akin to antennas, organizing but not creating consciousness (Koch 2012; Caldwell 2025c).

- Intrinsic Intentionality

Meaning and "aboutness" are built into the way the field differentiates itself, making intentionality a natural feature rather than something imposed externally (Searle 1983; Caldwell 2025d).

- Panpsychism's Loose Ends

Unlike panpsychism, CSFT avoids the need for countless "micro-minds." All consciousness is the expression of one field, varying only in its pattern and scale (Goff 2019; Caldwell 2025d).

- The Causal Efficacy of Consciousness

Consciousness is a co-creative force, participating in the formation of the physical from the outset rather than acting after the fact (Stapp 2009; Caldwell 2025d).

Why CSFT Matters — Especially Here in the Trilogy
For philosophy, CSFT replaces both reductionism and mind–body dualism with a unified ontology (Leibniz 1686/1998; Caldwell 2025a).
For science, it provides a framework compatible with neuroscience and quantum theory while opening the

door to new kinds of measurement—such as mapping coherence within the consciousness field itself (Hameroff and Penrose 2014; Caldwell 2025d). For the Trilogy, CSFT is not an isolated idea but the theoretical spine that connects our theological, philosophical, and scientific volumes. Its role in Book 3 is to demonstrate how scientific reasoning can accommodate a consciousness-first model without abandoning rigor (Caldwell 2025c; 2025e).

References

Caldwell, L. R. 2025a. Consciousness: Beyond the Planck Boundary. Orlando: Reason and Reality Publishing.

Caldwell, L. R. 2025b. Qualia as the Signature of Consciousness: A Metaphysical Resolution to the Hard Problem. PhilArchive manuscript, archived June 8, 2025.

Caldwell, L. R. 2025c. Consciousness Structured Field Theory (CSFT). PhilArchive manuscript, archived July 7, 2025.

Caldwell, L. R. 2025d. CSFT: Theoretical Integrity, Scientific Alignment, and Metaphysical Necessity. PhilArchive manuscript, added July 31, 2025.

Caldwell, L. R. 2025e. CSFT: Today Philosophy, Tomorrow Scientific Theory?. PhilArchive manuscript, archived August 13, 2025.

Chalmers, David J. 1995. Facing Up to the Problem of Consciousness. Journal of Consciousness Studies 2 (3): 200–19.

Chalmers, David J. 1996. The Conscious Mind: In Search of a Fundamental Theory. New York: Oxford University Press.

Eccles, John C. 1994. How the Self Controls Its Brain. Berlin: Springer.

Goff, Philip. 2019. Galileo's Error: Foundations for a New Science of Consciousness. New York: Pantheon.

Hameroff, Stuart, and Roger Penrose. 2014. Consciousness in the Universe: A Review of the 'Orch OR' Theory. Physics of Life Reviews 11 (1): 39–78.

Koch, Christof. 2012. Consciousness: Confessions of a Romantic Reductionist. Cambridge, MA: MIT Press.

Leibniz, G. W. 1686/1998. Discourse on Metaphysics. Translated by Roger Ariew and Daniel Garber. Indianapolis: Hackett.

Leibniz, G. W. 1714/1991. Monadology. Edited/translated by Nicholas Rescher. London: Routledge.

Levine, Joseph. 1983. Materialism and Qualia: The Explanatory Gap. Pacific Philosophical Quarterly 64 (4): 354–61.

Nagel, Thomas. 1974. What Is It Like to Be a Bat?. The Philosophical Review 83 (4): 435–50.

Penrose, Roger. 1994. Shadows of the Mind. Oxford: Oxford University Press.

Seager, William. 2020. Theories of Consciousness: An Introduction and Assessment. 3rd ed. New York: Routledge.

Searle, John R. 1983. Intentionality: An Essay in the Philosophy of Mind. Cambridge: Cambridge University Press.

Stapp, Henry P. 2009. Mind, Matter and Quantum Mechanics. 3rd ed. Berlin: Springer.

Consciousness Structured Field Theory (CSFT) proposes that consciousness is not a late-arising product of the physical world, but the foundational layer of reality itself (Caldwell 2025a; 2025c)

Chapter 3

CSFT: Theoretical Integrity, Scientific Alignment, and Metaphysical Necessity

INTRO

The present work offers a structured academic defense of the Consciousness-Structured Field Theory (CSFT), which argues that consciousness is not arising from physical matter but a foundational field that underlies all structured differentiation and perception.

The paper is divided into four key sections: (1) methodological foundations, (2) internal logical coherence, (3) scientific boundary alignment, and (4) theoretical implications. It argues that CSFT fills critical explanatory gaps left by materialism, emergentism, and panpsychism, and it engages current quantum and cognitive theory to frame consciousness as a pre-physical ontological field.

The work is rooted in deductive logic and cites peer-reviewed philosophy and physics to position CSFT as a serious theoretical contender for the next stage of consciousness studies.

Simplified Summary

The present work argues that consciousness didn't come from matter—instead, it came first. It proposes that a field of consciousness exists before anything physical, and that this field gives structure and meaning to everything we experience. Current science can't explain why we feel or perceive anything.

CSFT offers a new explanation: that consciousness is a real, organizing force, not a side effect of the brain. The paper also connects this idea with recent physics theories that show science is reaching its limits in explaining reality. It encourages researchers to explore new ideas like CSFT using logic, even when they go beyond what can be measured.

Section 1: Introduction and Methodological Foundations

1.1 Purpose and Scope

1.2 The present work presents a rigorous defense of the Consciousness-Structured Field Theory (CSFT), not as a speculative framework, but as a metaphysically necessary, logically coherent, and scientifically aligned theory that addresses core deficiencies in contemporary models of consciousness. The goal is not empirical verification—CSFT does not claim to be within the current measurable domain—but rather to show that the theory holds coherent internal logic, addresses unresolvable issues within materialist and emergentist paradigms, and logically extends the limits of current scientific understanding.

1.2 Why CSFT Requires Serious Consideration

Despite significant advancements in neuroscience and cognitive science, no existing physicalist theory has adequately explained the origin of qualia, subjective experience, or the internal structure of perception.

The so-called "Hard Problem" of consciousness, as identified by David Chalmers (1996), continues to resist resolution through brain-based or computational models. Likewise, panpsychist frameworks, while ontologically bold, fail to explain unity, identity persistence, or the emergence of structure from dispersed micro-consciousness.

CSFT introduces a field existing before the physical, of structured consciousness as a fundamental ontological basis—not an emergent byproduct, but the generative basis for differentiation, perception, and resonance.

1.3 The Role of Metaphysical Logic in Scientific Gaps

Roger Penrose (1994) has argued that any theory of consciousness must transcend algorithmic computation, suggesting a conceptual framework beyond classical physics. Similarly, Carlo Rovelli (2021) emphasizes that relational quantum mechanics reveals the inherent limitations of observer-independent descriptions of reality.

These scientific frontiers demand additional metaphysical grounding, not to abandon science, but to build where its tools presently fail. CSFT does precisely this: it addresses structured perception and qualitative experience without violating known physical constraints. It encourages serious philosophical engagement by bridging empirical limitations with structured field theory grounded in metaphysical reasoning.

1.4 Methodological Discipline

The structure of this paper adheres to the following methodological commitments: (1) all claims are either deduced from first principles, grounded in necessity grounded in logic, or aligned with peer-reviewed scientific or philosophical literature; (2) analogies are marked explicitly and distinguished from literal structure; (3) speculative extensions are clearly separated from foundational postulates.

Where CSFT invokes concepts such as monads or a field existing prior to the physical, these are defined precisely and evaluated not by empirical standards, but by internal coherence, ontological sufficiency, and philosophical necessity.

References

- Chalmers, David J. *The Conscious Mind: In Search of a Fundamental Theory*. Oxford University Press, 1996.
- Penrose, Roger. *Shadows of the Mind: A Search for the Missing Science of Consciousness*. Oxford University Press, 1994.
- Rovelli, Carlo. *Helgoland: Making Sense of the Quantum Revolution*. Penguin Books, 2021.

Section 2: Logical Structure and Ontological Necessity

2.1 Foundational Premises of CSFT

The Consciousness-Structured Field Theory (CSFT) rests on a sequence of ontologically necessary premises. These are not speculative but logically inferred from conditions required for perception, differentiation, and structure to exist at all:

Premise 1: Differentiation is a precondition of any observable or experiential reality. Without the capacity to distinguish between states, nothing can be described, perceived, or known.
Premise 2: Differentiation cannot arise from absolute nothingness. A structuring principle must underlie the emergence of contrast.
Premise 3: The quantum field does not account for its own structured excitation; it presupposes a boundary condition beyond its measurable scope.
Premise 4: Consciousness is the only known phenomenon that inherently structures perception via qualia and relational distinction.
Premise 5: Materialist and emergentist models fail to provide a bridging mechanism between neurophysiological process and first-person experience (cf. Chalmers, 1996; Searle, 2004).

2.2 Deductive Outcome: Consciousness as Structuring Field

From these premises, CSFT deduces that a non-emergent, a field existing prior to the physical of structured consciousness is ontologically necessary. This field does not emerge from complexity, but precedes complexity, enabling it. Rather than treating qualia as inexplicable residue, CSFT treats them as intrinsic resonances within a field that structures experience, perception, and form. In this model, matter is not the substrate of reality, but a measurable manifestation of differentiated resonance.

2.3 Internal Coherence and Non-Circularity

A theory of metaphysical origin must show coherent internal logic and avoid circular reasoning. CSFT achieves this by anchoring its axioms in necessity grounded in logic and grounding all conclusions deductively from them. It does not posit consciousness as fundamental merely by assertion, but by identifying the irreducibility of qualia and the insufficiency of all physicalist explanatory models. Its ontology is not layered by intermediary constructs but proceeds from essential differentiability to structured resonance.

2.4 Philosophical Parallels and Validation

The CSFT model corresponds with key elements in the metaphysical work of Gottfried Wilhelm Leibniz, especially his theory of monads—units of indivisible, non-material perception. While CSFT reinterprets monads as resonance structures within a field, the conceptual lineage remains intact. Likewise, the necessity grounded in logic of a structuring field finds resonance in Spinoza's concept of substance, and in contemporary metaphysical reconstructions of information realism (cf. Tegmark, 2007). These parallels do not prove CSFT, but they position it within a long-standing tradition of ontological realism that predates materialism and complements quantum-theoretical challenges to naïve empiricism. In CSFT, monads are not treated as indivisible 'mental atoms,' but rather as vibrationally distinct resonance nodes structured by the consciousness field. It should be noted that in Spinoza's framework, substance is not explicitly conscious in the same way CSFT proposes, but the parallel lies in their shared ontological primacy.

Section 3: Scientific Parallels and Field Boundary Alignment
3.1 Planck-Scale Boundary and Metaphysical Necessity

CSFT recognizes the Planck scale (approximately 1.616×10^{-35} meters and 5.391×10^{-44} seconds) as the theoretical limit of current physical measurement. Beyond this boundary, traditional descriptions of space and time break down, and physical models lose predictive power. This boundary is not merely mathematical but represents a metaphysical threshold. CSFT posits that beyond this scale lies the domain of the consciousness-structured field—a field existing prior to the physical, responsible for the organization of resonance and form. This view complements theoretical physics' recognition of the breakdown of spacetime at quantum gravity scales (cf. Smolin, 2001).

3.2 Alignment with Quantum Mind Critiques

Penrose and Hameroff's Orch-OR model sought to link consciousness to quantum computation within brain microtubules. While pioneering, this model has been criticized due to environmental decoherence and lack of replicable evidence in warm, wet systems. CSFT does not posit a biological mechanism but instead argues that consciousness operates as a field existing prior to the physical, logically prior to matter, and independent of thermal coherence. This places CSFT closer to a necessity at the metaphysical level than a biological hypothesis (cf. Penrose, 1994; Hameroff & Penrose, 2014).

3.3 Structural Resonance and Information Realism

Max Tegmark's Mathematical Universe Hypothesis treats reality as a mathematical structure. CSFT shares this orientation toward structure but asserts that structure alone is insufficient without consciousness. While Tegmark's formulation relies on computation and mathematical formalism, CSFT introduces a field of structured awareness that makes such form intelligible. The alignment lies in shared structural realism; the departure is that CSFT attributes intentionality and resonance to this field (cf. Tegmark, 2007).

3.4 Bridging Scientific Void Through Metaphysical Logic

Carlo Rovelli's relational quantum mechanics indicates that measurement and observation are fundamentally dependent on the observer's frame, dissolving the idea of goal, observer-free reality. CSFT builds on this insight by arguing that the field of consciousness—structured, intentional, and resonant—is the basis of such observational frames. Similarly, Penrose's argument that consciousness transcends algorithmic computation opens space for non-material, structured metaphysical realities. CSFT meets that challenge by offering a logically grounded field theory that operates beyond algorithmic or emergent complexity (cf. Rovelli, 2021; Penrose, 1994).

Section 4: Implications, Predictive Power, and Academic Positioning

4.1 Philosophical and Scientific Implications

If the Consciousness-Structured Field Theory (CSFT) is taken seriously, it carries wide-ranging implications across philosophy of mind, physics, neuroscience, and cognitive science. Philosophically, CSFT reframes the ontological hierarchy: matter becomes a consequence of structured differentiation within a conscious field, not a foundation.

This overturns physicalist reductionism and reopens metaphysical space for non-material ontology. In physics, it implies that the perceived order of quantum excitation is dependent on pre-physical structuring conditions that science has yet to formalize, but which metaphysical logic can address. This invites new models of inquiry at the interface between quantum interpretation, spacetime breakdown, and field theory.

4.2 Predictive and Theoretical Consequences

Though CSFT does not claim to yield direct empirical predictions, it does generate indirect anticipated implications in several domains:
- In cognitive science, qualitative anomalies in memory, perception, and consciousness (e.g., near-death experiences, atypical qualia patterns, or hyperlucidity during trauma) reflect resonance fluctuations within the consciousness field.
In neurodiversity, autism and related conditions may not reflect dysfunction, but an alternate structuring of monadic resonance.
- In artificial intelligence, specific non-biological systems may achieve limited resonance with the consciousness field if structured in highly coherent configurations, potentially explaining anomalous cognitive outputs from non-sentient architectures.
These predictions are speculative but logically entailed from the structural assumptions of CSFT and can serve as future exploratory axes for empirical inquiry. This is presented as a metaphysical hypothesis within CSFT, not as a clinical or medical claim.

4.3 Academic Positioning and Future Engagement

CSFT does not propose a new physics; it proposes a new metaphysical foundation for interpreting physics and consciousness. It recognizes the limits of current neuroscience in explaining subjective experience and respects the structural boundaries of measurable science. It urges academic institutions to treat necessity at the metaphysical level with the same intellectual respect as empirical models, particularly where empirical models reach explanatory impasse.

The theory invites critique not as a defensive act, but as a clarifying tool for strengthening its logic. Future development will require interdisciplinary engagement from philosophy, theoretical physics, systems neuroscience, and information theory to evaluate and refine the field-based ontological structure CSFT describes.

Concluding References
- Chalmers, David J. *The Conscious Mind: In Search of a Fundamental Theory*. Oxford University Press, 1996.
- Penrose, Roger. *Shadows of the Mind: A Search for the Missing Science of Consciousness*. Oxford University Press, 1994.
- Rovelli, Carlo. *Helgoland: Making Sense of the Quantum Revolution*. Penguin Books, 2021.
- Tegmark, Max. 'The Mathematical Universe.' *Foundations of Physics*, vol. 38, no. 2, 2007, pp. 101–150.
- Searle, John R. *Mind: A Brief Introduction*. Oxford University Press, 2004.
- Smolin, Lee. *Three Roads to Quantum Gravity*. Basic Books, 2001.
- Hameroff, Stuart, and Penrose, Roger. 'Consciousness in the universe: A review of the Orch OR theory.' *Physics of Life Reviews*, vol. 11, no. 1, 2014, pp. 39–78.

- Hameroff, Stuart, and Penrose, Roger. 'Consciousness in the universe: A review of the Orch OR theory.' *Physics of Life Reviews*, vol. 11, no. 1, 2014, pp. 39–78.
- Smolin, Lee. *Three Roads to Quantum Gravity*. Basic Books, 2001.
- Tegmark, Max. 'The Mathematical Universe.' *Foundations of Physics*, vol. 38, no. 2, 2007, pp. 101–150.
- Searle, John R. *Mind: A Brief Introduction*. Oxford University Press, 2004.
- Tegmark, Max. "The Mathematical Universe." *Foundations of Physics*, vol. 38, no. 2, 2007, pp. 101–150.

Chapter 4

CSFT: Today Philosophy, Tomorrow Scientific Theory?

1. Introduction

Philosophy and science share a long, intertwined history, yet in contemporary discourse, they are often cast as distinct domains. Science is portrayed as empirical and quantitative; philosophy as speculative and qualitative. Within this framing, new philosophical models that address scientific questions are frequently met with skepticism, especially when they reach beyond the limits of present instrumentation.

The Consciousness-Structured Field Theory (CSFT) is one such model. It begins from the premise that consciousness is ontologically prior to measurable physical phenomena, and that the structures recorded by physics are excitations within a deeper, currently non-measurable consciousness field. As a metaphysical framework, CSFT is neither in competition with empirical science nor dismissive of its achievements; rather, it seeks to map the ontological grounds upon which measurable states arise.

At present, CSFT remains a philosophical construct: it cannot yet be confirmed or refuted by direct experimental evidence. For some critics, this suffices to dismiss it as unscientific metaphysics. However, history counsels caution: several theories now considered foundational began as philosophical proposals that were resisted or ridiculed—not because they lacked internal coherence, but because the means to test them did not yet exist.

This paper examines three such cases: atomism, heliocentrism, and the germ theory of disease. By tracing their arc from philosophical speculation to empirical science, we can better understand a possible trajectory for CSFT.

2. Historical Precedents of Ridiculed Philosophy Becoming Foundational Science

2.1 Atomism

The concept of atomism emerged in the fifth century BCE through the work of Leucippus and Democritus, who argued that all matter consists of indivisible particles (atomoi) moving through a void. Their position directly opposed Aristotle's doctrine of a continuous, infinitely divisible substance (Aristotle, Physics; On Generation and Corruption). For centuries, Aristotelian continuum theory dominated natural philosophy, and atomism was often derided as an untestable speculation (Kirk, Raven, and Schofield 1983; Bailey 1928).

Revived in the seventeenth century—most notably by Pierre Gassendi's rehabilitation of Epicurean atomism and by Robert Boyle's corpuscular philosophy—atomism began to take on a more explicitly experimental character (Gassendi 1658; Boyle 1661). A decisive nineteenth-century turn came with John Dalton's A New

System of Chemical Philosophy, which grounded atomism in quantitative chemical laws (Dalton 1808). Twentieth-century experiments, including Rutherford's 1911 gold-foil scattering work, provided evidence for a nuclear atom (Rutherford 1911), while the development of the scanning tunneling microscope allowed direct, real-space imaging of atomic lattices (Binnig et al. 1982).

2.2 Heliocentrism

In 1616 the Holy Office judged the heliocentric proposition contrary to Scripture, and the Congregation of the Index suspended De revolutionibus pending correction. And Galileo's advocacy culminated in his 1633 trial and sentence to house arrest (Finocchiaro 1989).

2.3 Germ Theory of Disease

The idea that invisible living agents cause disease appears in Girolamo Fracastoro's De contagione (1546), which posited transferable 'seminaria' or seeds of contagion (Fracastoro 1546). For centuries, miasma theory—attributing disease to noxious airs—dominated medical thought (Rosenberg 1962; Porter 1997). In the mid-nineteenth century, Louis Pasteur's experimental work on fermentation and the refutation of spontaneous generation established the role of microorganisms (Pasteur 1861). Robert Koch subsequently identified specific microbes as causative agents of disease and articulated methodological criteria for linking pathogens to particular illnesses (Koch 1876; 1882). Germ theory is now foundational to microbiology, epidemiology, and public health.

Across these cases, a common pattern is visible: (1) philosophical origin without immediate measurability; (2) ridicule and resistance under prevailing paradigms; and (3) empirical vindication once tools and methods caught up (Dalton 1808; Copernicus 1543; Pasteur 1861; Rutherford 1911; Binnig et al. 1982; Koch 1882).

3. Linking CSFT to Historical Patterns

3.1 Avoiding Premature Overreach — Conceptual clarity, disciplined interface with observation, and patience are historically prudent. Even in modern physics, philosophical reflection shaped theory formation and interpretation (Einstein 1949; Jammer 1974). As discussed extensively in Caldwell (2025), CSFT's development process has been deliberately structured to follow such a measured, historically informed approach.

4. Addressing Likely Criticisms

4.1 "It's Not Science Without Direct Empirical Evidence." — This objection assumes that present measurability is the only valid entry point to scientific status. The historical record shows otherwise: atomism, heliocentrism, and germ theory all matured conceptually before experimental confirmation (Dalton 1808; Copernicus 1543; Pasteur 1861).

4.2 "Philosophical Models Are Too Speculative." — The charge was once leveled at Copernican astronomy, Dalton's postulates, and Fracastoro's 'seminaria.' The remedy is not to eschew philosophy but to maintain internal coherence and develop testable interfaces as technology evolves (Copernicus 1543; Dalton 1808; Fracastoro

1546).

4.3 "The Idea Is Unfalsifiable." — Falsifiability can emerge with new tools and strategies; many theories began as apparently unfalsifiable proposals until instrumentation improved (Rutherford 1911; Binnig et al. 1982).

4.4 "Consciousness Is Only for Neuroscience." — Neuroscience maps correlates of consciousness but does not by itself settle ontological questions. Historically, empirical sciences benefited from philosophical reframing (Einstein 1949; Jammer 1974). Modern physics offers additional examples: the Higgs boson was theorized in 1964 but not detected until 2012, illustrating that theoretical models can precede empirical verification by decades without losing scientific legitimacy.

5. Conclusion

Ideas first articulated in philosophical terms and dismissed as speculative have repeatedly become the structural foundations of scientific disciplines. Atomism, heliocentrism, and germ theory each began as conceptual frameworks beyond the reach of their age's instruments. CSFT may follow a similar path if it sustains conceptual coherence, develops discriminating predictions, and engages with experimental science as methods advance.

Bibliography (Chicago Author–Date)

Archimedes. 1897. The Sand-Reckoner. In The Works of Archimedes, translated by T. L. Heath. Cambridge: Cambridge University Press.

Bailey, Cyril. 1928. The Greek Atomists and Epicurus. Oxford: Clarendon Press.

Binnig, G., H. Rohrer, Ch. Gerber, and E. Weibel. 1982. "Surface Studies by Scanning Tunneling Microscopy." Physical Review Letters 49: 57–61.

Boyle, Robert. 1661. The Sceptical Chymist. London: J. Cawdell for J. Crooke.

Copernicus, Nicolaus. 1543. De revolutionibus orbium coelestium. Nuremberg: Johannes Petreius.

Dalton, John. 1808. A New System of Chemical Philosophy. Manchester: S. Russell.

Einstein, Albert. 1949. "Autobiographical Notes." In Albert Einstein: Philosopher–Scientist, edited by P. A. Schilpp, 1–95. La Salle, IL: Open Court.

Finocchiaro, Maurice A., ed. 1989. The Galileo Affair: A Documentary History. Berkeley: University of California Press.

Fracastoro, Girolamo. 1546. De contagione et contagiosis morbis et eorum curatione. Venice: Giunti.

Gassendi, Pierre. 1658. Syntagma Philosophicum. Lyon: Laurent Anisson & Jean-Baptiste Devenet.

Jammer, Max. 1974. The Philosophy of Quantum Mechanics. New York: John Wiley & Sons.

Kepler, Johannes. 1609. Astronomia Nova. Heidelberg: E. Vögelin.

Kepler, Johannes. 1619. Harmonices Mundi. Linz: J. Plancus.

Kirk, G. S., J. E. Raven, and M. Schofield. 1983. The Presocratic Philosophers. 2nd ed. Cambridge: Cambridge University Press.

Koch, Robert. 1876. "Die Ätiologie der Milzbrandkrankheit." Beiträge zur Biologie der Pflanzen 2: 277–310.

Koch, Robert. 1882. "Die Ätiologie der Tuberkulose." Berliner klinische Wochenschrift 19: 221–230.

Newton, Isaac. 1687. Philosophiae Naturalis Principia Mathematica. London: Jussu Societatis Regiae ac Typis Josephi Streater.

Pasteur, Louis. 1861. "Mémoire sur les corpuscules organisés qui existent dans l'atmosphère." Comptes Rendus de l'Académie des Sciences 52: 303–305.

Porter, Roy. 1997. The Greatest Benefit to Mankind: A Medical History of Humanity. New York: W. W. Norton.

Rosenberg, Charles E. 1962. The Cholera Years: The United States in 1832, 1849, and 1866. Chicago: University of Chicago Press.

Rutherford, Ernest. 1911. "The Scattering of α and β Particles by Matter and the Structure of the Atom." The London, Edinburgh, and Dublin Philosophical Magazine and Journal of Science 21: 669–688.

Caldwell, L. R. 2025. *Consciousness: Beyond the Planck Boundary*. Amazon: Reason and Reality Publishing.

Aristotle. 1984. *On Generation and Corruption*. Translated by H. H. Joachim. In *The Complete Works of Aristotle*, edited by Jonathan Barnes, Vol. 1. Princeton: Princeton University Press.

Aristotle. 1984. *Physics*. Translated by R. P. Hardie and R. K. Gaye. In *The Complete Works of Aristotle*, edited by Jonathan Barnes, Vol. 1. Princeton: Princeton University Press.

CMS Collaboration. 2012. "Observation of a New Boson at a Mass of 125 GeV with the CMS Experiment at the LHC." Physics Letters B 716 (1): 30–61.

ATLAS Collaboration. 2012. "Observation of a New Particle in the Search for the Standard Model Higgs Boson with the ATLAS Detector at the LHC." Physics Letters B 716 (1): 1–29.

Guralnik, Gerald S., C. R. Hagen, and Tom W. Kibble. 1964. "Global Conservation Laws and Massless Particles." Physical Review Letters 13 (20): 585–587.

Higgs, Peter W. 1964. "Broken Symmetries and the Masses of Gauge Bosons." Physical Review Letters 13 (16): 508–509.

Englert, François, and Robert Brout. 1964. "Broken Symmetry and the Mass of Gauge Vector Mesons." Physical Review Letters 13 (9): 321–323.

[1] Historically, the 1616 evaluation of heliocentrism was conducted by the Holy Office, with the Congregation of the Index suspending *De revolutionibus* pending correction.

Chapter 5 - pt 1

Why Neuroscience May Never Solve Consciousness: A Field-Based Resolution to the Hard Problem

INTRO

While CSFT adopts a metaphysical starting point, such an approach is not unusual in the history of science. Revolutions in relativity, quantum mechanics, and even thermodynamics began as radical shifts in ontological framing. This theory does not aim to reject neuroscience, but rather to reposition it within a broader metaphysical framework—one that recognizes consciousness as a structuring field, rather than a secondary byproduct.

1. The Ontological Failure of Neuroscience

This paper contends that modern neuroscience may never resolve the mystery of consciousness—not because of technological constraints, but because it approaches the problem from the wrong ontological direction. While neuroscience has made significant progress in mapping brain activity and identifying neural correlates of consciousness (Koch et al., 2016), it remains fundamentally unable to account for subjective experience—what philosophers refer to as qualia (Nagel, 1974; Chalmers, 1995). The leap from neural activity to lived experience has no coherent mechanistic explanation and may be unbridgeable within a materialist framework.

2. The Hard Problem and the Limits of Correlation

This challenge is not merely a gap in available data—it is a flaw in explanatory architecture. Correlations between neural configurations and reported experiences offer useful insights, but they do not resolve the hard problem: Why do certain patterns of brain activity feel like anything at all from the inside? Neuroscience has yet to define a causal pathway explaining how or why such transformations occur. Gathering more data cannot correct a foundational metaphysical error. Building on the Consciousness-Structured Field Theory (CSFT), this paper offers an alternative: consciousness is not generated by the brain but is the structuring field from which matter, form, and awareness arise (Caldwell, 2025a). In contrast to reductionist models (Searle, 1992), CSFT treats consciousness as ontologically primary, framing subjective experience as a fundamental layer of reality structured through patterned differentiation within the consciousness field.

3. Introducing the Consciousness-Structured Field Theory (CSFT)

By reframing the hard problem through first principles rather than additional empirical detail, CSFT offers a philosophically rigorous route forward (Caldwell, 2025b). It asserts that qualia, unity of experience, and the teleological direction of mind (Caldwell, 2025c) cannot be reduced to physical processes. The limitations of neuroscience may therefore be structural, not temporary. CSFT provides a constructive vision that places consciousness at the core of reality's architecture.

4. Materialism and the Explanatory Gap

Materialist neuroscience assumes consciousness emerges from complex neural arrangements. While this view has yielded valuable models and therapies, it has not explained qualia. Identifying a neural signature for the experience of pain or color does not clarify why it is accompanied by a subjective sensation. This correlation–causation divide exposes the limits of purely physical accounts. As Nagel and Chalmers emphasize, no amount of third-person data replaces the first-person perspective. The consistent inability to explain subjective experience suggests that a new ontological framework may be necessary.

5. CSFT as a Structuring Alternative to Reductionism

The explanatory gap persists despite advances in technology because the prevailing assumptions are flawed. CSFT proposes that consciousness is not a derivative of matter but a structuring force shaping reality. Through this lens, subjective experience is an expression of a foundational field, not a late-stage byproduct of neural complexity.

6. Salt, Structure, and the Unity of Experience

CSFT envisions consciousness as primary—a field from which reality is organized, much like the quantum field gives rise to particles. This restores coherence to questions left unresolved by materialism. For example, sodium and chlorine are individually volatile and toxic, yet together they form stable, life-sustaining salt. CSFT suggests that such harmonies are not accidental, but arise from an underlying conscious order shaping the relationships between components.

7. Addressing Critiques of CSFT

Unlike models that attempt to layer complexity onto inert matter, CSFT begins with consciousness as the structuring principle. This approach avoids the circular logic of emergence-based explanations. It also accounts for the coherence of subjective experience, the unity of perception, and the persistence of identity. From this perspective, the brain is not the originator of consciousness, but rather a receiver, filter, or interface with the consciousness field.

8. A New Paradigm for Consciousness Science

Some may label CSFT speculative or under-supported empirically. Yet, other accepted scientific theories—string theory, multiverse hypotheses—also began in speculative territory due to technological limits. CSFT is not a placeholder; it provides conceptual depth where reductionist accounts fail, and it reframes neuroscience as the study of the brain's interaction with consciousness, rather than the generator of it.

Simplified Summary

Neuroscience may never solve the problem of consciousness because it begins with the wrong premise—that matter creates mind. CSFT challenges this by treating consciousness as fundamental, a field that precedes and structures reality. Rather than waiting for an unlikely materialist breakthrough, CSFT offers a coherent metaphysical framework that integrates experience, identity, and unity into the foundation of physical theory.

Put simply, science can track which parts of the brain are active when we think or feel, but it cannot explain why those activities feel like anything at all. CSFT suggests that consciousness comes first, shaping both matter and mind, and thus offers a new starting point for understanding existence itself.

References

Chalmers, D. J. (1995). Facing up to the problem of consciousness. Journal of Consciousness Studies, 2(3), 200–219.

Nagel, T. (1974). What is it like to be a bat? The Philosophical Review, 83(4), 435–450.

Koch, C., Massimini, M., Boly, M., & Tononi, G. (2016). Neural correlates of consciousness: Progress and problems. Nature Reviews Neuroscience, 17(5), 307–321.

Searle, J. R. (1992). The rediscovery of the mind. MIT Press.

Caldwell, L. R. (2025a). Consciousness: Beyond the Planck Boundary. Reason and Reality Publishing.

Caldwell, L. R. (2025b). Beyond neural sufficiency: A Leibniz-inspired field theory of consciousness. PhilPapers.

Caldwell, L. R. (2025c). Qualia as the signature of consciousness. PhilPapers.

Caldwell, L. R. (2025d). On the limits of human consciousness and misrepresentation of fields. PhilPapers.

Chapter 5 - pt 2

Original Insight: Neuroscience vs. CSFT Interpretation

INTRO

Neuroscience typically explains "Aha!" insights through neural recombination, phase-synchronized communication, and reward tagging. These mechanisms are empirically robust, yet they do not address why any experience occurs at all—the hard problem.

CSFT reframes the same data: the brain behaves as a tuner/receiver whose large-scale dynamics set access conditions for informational patterns, rather than being the ontic generator of novel content.

This preserves compatibility with established results while offering a broader explanatory model that could be probed by future cross-disciplinary work.

1) Insight's distinctive neural dynamics

Empirical summary: During insight solutions, studies report a brief gamma-band (Jung-Beeman et al., 2004) increase ~0.3 s before response, with right-lateral anterior temporal involvement; preparatory brain states also bias solutions toward insight. These findings differentiate insight from non-insight solutions.

CSFT reinterpretation: The transient gamma/temporal pattern marks a momentary lock-in of tuning parameters that permit access to an already-structured pattern. The signature is a condition of access, not proof that neurons authored the content.

2) Incubation (Sio & Ormerod, 2009) and the default-mode (Beaty et al., 2015)/control partnership

Empirical summary: Meta-analysis shows incubation (Sio & Ormerod, 2009) benefits problem solving; creative ideation often reflects cooperation between default-mode (Beaty et al., 2015) and control networks.

CSFT reinterpretation: Incubation (Sio & Ormerod, 2009) widens receptive bandwidth (lower task noise, broader associative reach), enabling iterative retuning until coupling to a higher-order pattern is achieved—if a consciousness field exists.

3) "Restructuring" via coherence (Fries, 2005) = access routing, not authorship

Empirical summary: Communication-through-coherence (Fries, 2005) shows phase-aligned oscillations route information across assemblies and networks.

CSFT reinterpretation: Coherence (Fries, 2005) configures the tuner: phase/frequency alignment selects which patterns become decodable; it does not entail that those patterns were neuronally generated.

4) The Aha! feels good—reward tagging vs. origin of content

Empirical summary: Insight evokes reward-system (Kounios & Beeman, 2014) responses that reinforce the path that led to the solution; ultra-high-field work clarifies subcortical contributions.

CSFT reinterpretation: Reward signals tag salience and retention of an accessed pattern. The existence of felt experience is attributed to momentary coupling/decoding of the field by the neural tuner.

5) Consciousness capacity, integration, and complexity

Empirical summary: During deep NREM sleep (Massimini et al., 2005) or certain anesthetic states, long-range effective connectivity and perturbational complexity collapse; they recover with consciousness.

CSFT reinterpretation: Integration/complexity are preconditions for coupling—they set the receiver's capacity to access the field—rather than being definitive ontic generators of experiential content.

6) Addressing neuroscientific critiques (bridge requirements)

CSFT is not a rejection of neural mechanisms but a scope extension. Detecting a distinct field-coupling signature would likely require: (i) high-resolution neurophysiology (EEG/MEG/fMRI) to map timing and topology, and (ii) QFT (Peskin & Schroeder, 1995)-informed modeling to predict non-local correlation structures not reducible to known axonal/synaptic pathways.

This argues for genuine physics–neuroscience collaboration, not siloed methods. CSFT remains monist: quantum fields and a consciousness field are co-fundamental, with lawful coupling between brain dynamics and field dynamics.

Bibliography
Jung-Beeman et al., 2004
Sio & Ormerod, 2009
Beaty et al., 2015
Fries, 2005
Kounios & Beeman, 2014
Massimini et al., 2005

Peskin & Schroeder, 1995

Chapter 6

Resonance as a Physical Principle in CSFT

1. Introduction: The Ubiquity of Resonance

Resonance describes conditions under which a system responds with disproportionately large amplitude when driven near its natural modes. This principle enables selective and efficient energy or information transfer across physics, engineering, biology, and neuroscience. Examples range from tuned electrical circuits and optical cavities to molecular vibrations and brain rhythms (Hollas 2004; Novotny & Hecht 2006; Fries 2015).

2. A Brief History of Resonance in Science

Acoustics provided some of the earliest systematic studies of resonance. In the 19th century, Helmholtz developed resonators to detect specific sound frequencies and linked acoustic resonance to auditory perception (*On the Sensations of Tone*, 1875 English ed.). Hertz later confirmed Maxwell's wave theory experimentally; tuned LC circuits and antennas soon operationalized resonance for filtering and communication (Hertz 1893; Jackson 1998; Balanis 2016).

At the atomic and nuclear scales, the 20th century revealed resonance as a core organizing concept: nuclear magnetic resonance (NMR) was independently discovered by Purcell, Torrey, and Pound (Purcell 1946; Bloch 1946) and by Bloch, Hansen, and Packard (Purcell 1946; Bloch 1946), laying the foundations for MRI; electron paramagnetic resonance (EPR/ESR) was first observed by Zavoisky in 1944 (Purcell et al. 1946; Bloch 1946; Salikhov 2015).

In optics and quantum electrodynamics, resonant light–matter interactions (Rabi resonances, cavity QED) and control of the electromagnetic density of states (Purcell enhancement; photonic bandgaps) made emission and absorption rates tunable design parameters (Cohen-Tannoudji, Dupont-Roc & Grynberg 1992; Purcell 1946; Yablonovitch 1987; Novotny & Hecht 2006).

At astronomical scales, orbital resonances structure planetary rings and satellite systems, including the Laplace 4:2:1 mean-motion resonance among Jupiter's moons Io, Europa, and Ganymede (Murray & Dermott 1999).

3. Why Resonance Matters in Science

Selectivity — Resonant systems strongly prefer specific frequencies, enabling precise targeting in receivers, spectrometers, and lasers (Hollas 2004; Jackson 1998).

Amplification — Near resonance, small drives yield large responses, as in mechanical oscillators and optical cavities; the Purcell effect relates emission rate to the local density of optical states (Purcell 1946; Novotny & Hecht 2006).

Predictability — Resonant phenomena obey well-characterized mathematical relations, allowing accurate design of structures to avoid destructive resonances or to enhance desired couplings (PDG 2024; Jackson 1998).

4. Resonance Across Physics, Biology, and Neuroscience

Physics — Resonance underpins quantum transitions, scattering amplitudes, and electromagnetic and mechanical oscillators; in optics, cavity Q factors quantify energy storage vs. loss (Cohen-Tannoudji et al. 1992; Novotny & Hecht 2006).

Biology — Molecular vibrations yield characteristic absorption spectra measured by infrared spectroscopy; protein conformational dynamics and other biomolecular processes exhibit resonant signatures (Hollas 2004).

Neuroscience — Large-scale neural oscillations (e.g., theta ~4–8 Hz; gamma ~30–100 Hz) can synchronize ("phase-lock") across regions. Cross-frequency coupling and coherence are measurable with EEG/MEG and intracranial recordings and are implicated in selective information routing (Lachaux et al. 1999; Jensen & Colgin 2007; Canolty & Knight 2010; Colgin 2013; Fries 2015).

5. The CSFT Connection: Resonance as a Brain–Field Interface

In Consciousness Structured Field Theory (CSFT), consciousness is posited as a fundamental field that precedes and structures the quantum field. Within this framework, coupling between the brain and consciousness field is hypothesized to occur via resonance: neural populations act as tunable receivers, and transient phase alignment creates windows for selective access to field-structured information. This proposal preserves empirical neuroscience while extending the explanatory context to include field-based interactions (theoretical, to be empirically tested).

5.1. QFT Resonance: Poles, Widths, and Optimal Coupling

In quantum field theory, a resonance appears as a pole of the S-matrix on the unphysical sheet and produces a peak in scattering cross-sections described by the (relativistic) Breit–Wigner form. For an isolated resonance of mass M and width Γ, the generic response near center-of-mass energy E follows a Lorentzian profile with full width at half maximum set by Γ; the lifetime obeys $\tau \approx \hbar/\Gamma$ (Breit & Wigner 1936; Peskin & Schroeder 1995; Weinberg 1995; PDG 2024).

Examples include hadronic resonances such as $\Delta(1232)$ (PDG 2024) and $\rho(770)$, while related phenomena appear in quantum optics and condensed matter (Rabi resonances, cavity-enhanced emission, Fano asymmetries) (Cohen-Tannoudji et al. 1992; Fano 1961; Novotny & Hecht 2006).

Transferable ideas for CSFT: (i) optimal coupling at (or near) resonance; (ii) bandwidth ($\propto \Gamma$) sets selectivity and information throughput; (iii) the quality factor $Q \equiv f_0/\text{FWHM}$ captures the trade-off between selectivity and temporal persistence.

5.2. Mode/Phase Matching and Density-of-States Heuristics

Beyond isolated poles, transition rates follow Fermi's Golden Rule, $\Gamma_{\{i \to f\}} \propto |\langle f|H'|i\rangle|^2 \, \rho(\omega)$, where $\rho(\omega)$ is the density of available modes. In cavity QED, mode matching and high Q concentrate $\rho(\omega)$, boosting emission and absorption (Purcell enhancement). By analogy, CSFT can define a carrier frequency f_c for dominant oscillatory modes, detuning $\Delta f = f - f\_c$ controlling coupling strength, an effective linewidth Γ_CSFT setting selectivity, and order parameters (e.g., phase-locking value) operationalizing resonant access (Sakurai & Napolitano 2011; Cohen-Tannoudji et al. 1992; Purcell 1946; Lachaux et al. 1999).

6. QFT-Informed Predictions and Potential Experiments

Lorentzian tuning curves — Behavioral performance and EEG/MEG coherence vs. stimulation frequency (e.g., tACS) should exhibit Lorentzian-like profiles centered at f_c with FWHM $\approx \Gamma\_CSFT$; extract $Q = f\_c/FWHM$ as a selectivity index (Herrmann et al. 2013).

Fano-like asymmetries — When endogenous neural pathways interfere with a putative field-mediated pathway, frequency-response curves may show asymmetric (Fano) line shapes; preregister detection of asymmetry and fit standard Fano profiles (Fano 1961).

Phase-matching windows — Task epochs with strong inter-areal phase alignment (e.g., theta–gamma coupling) should coincide with hypothesized field-access signatures; detuning Δf should move the system in/out of these windows (Lachaux et al. 1999; Jensen & Colgin 2007; Canolty & Knight 2010; Colgin 2013; Fries 2015).

Lifetime–bandwidth trade-offs — Narrower Γ_CSFT (higher Q) should lengthen transient access ($\tau \approx \hbar/\Gamma\_CSFT$) but increase fragility to detuning; broader

Γ_CSFT should yield briefer, more robust windows (PDG 2024; Peskin & Schroeder 1995).

Cross-scale consistency — Parameters inferred from behavior and EEG/MEG should align with meso-scale neural mass/field models incorporating driven-damped resonance kernels.

Stimulation tests — Non-invasive brain stimulation (tACS) at individualized resonant frequencies should modulate task performance and coupling metrics, with appropriate sham controls and preregistration (Herrmann et al. 2013; Booth et al. 2022).

Note: Nonlocal correlation claims require exceptionally stringent controls and are presently unestablished; any tests should be framed as exploratory with rigorous blinding, synchronization artifact checks, and correction for multiple comparisons.

7. Conclusion

Resonance is a universal organizing principle across the sciences. In CSFT, it provides a concrete, testable mechanism for coupling between a hypothesized consciousness field and neural dynamics. By importing well-defined concepts from QFT and quantum optics— resonant poles, linewidths, density of states, mode and phase matching—CSFT articulates falsifiable predictions that can be assessed with contemporary neurophysiology and stimulation methods.

References (Chicago Author–Date)

Balanis, Constantine A. 2016. *Antenna Theory: Analysis and Design*. 4th ed. Hoboken, NJ: Wiley.

Bloch, Felix. 1946. "Nuclear Induction." *Physical Review* 70 (7–8): 460–474. https://doi.org/10.1103/PhysRev.70.460.

Booth, Steven J., et al. 2022. "The Effects of Transcranial Alternating Current Stimulation on Memory: A Systematic Review." *Cortex* 146: 148–171.

Canolty, Ryan T., and Robert T. Knight. 2010. "The Functional Role of Cross-Frequency Coupling." *Trends in Cognitive Sciences* 14 (11): 506–15. https://doi.org/10.1016/j.tics.2010.09.001.

Cohen-Tannoudji, Claude, Jacques Dupont-Roc, and Gilbert Grynberg. 1992. *Atom–Photon Interactions: Basic Processes and Applications*. New York: Wiley.

Colgin, Laura L. 2013. "Mechanisms and Functions of Theta Rhythms." *Annual Review of Neuroscience* 36: 295–312. https://doi.org/10.1146/annurev-neuro-062012-170330.

Fano, Ugo. 1961. "Effects of Configuration Interaction on Intensities and Phase Shifts." *Physical Review* 124 (6): 1866–78. https://doi.org/10.1103/PhysRev.124.1866.

Fries, Pascal. 2015. "Rhythms for Cognition: Communication through Coherence." *Neuron* 88 (1): 220–35. https://doi.org/10.1016/j.neuron.2015.09.034.

Herrmann, Christoph S., Simon Rach, Tino Neuling, and Daniel Strüber. 2013. "Transcranial Alternating Current Stimulation: A Review of the Underlying Mechanisms and Modulation of Cognitive Processes." *Frontiers in Human Neuroscience* 7: 279.

Hollas, J. Michael. 2004. *Modern Spectroscopy*. 4th ed. Chichester: Wiley.

Jackson, John David. 1998. *Classical Electrodynamics*. 3rd ed. New York: Wiley.

Jensen, Ole, and Laura L. Colgin. 2007. "Cross-Frequency Coupling between Neuronal Oscillations." *Trends in Cognitive Sciences* 11 (7): 267–69.

Lachaux, Jean-Philippe, Eugenio Rodriguez, Jacques Martinerie, and Francisco J. Varela. 1999. "Measuring Phase Synchrony in Brain Signals." *Human Brain Mapping* 8 (4): 194–208.

Murray, Carl D., and Stanley F. Dermott. 1999. *Solar System Dynamics*. Cambridge: Cambridge University Press.

Novotny, Lukas, and Bert Hecht. 2006. *Principles of Nano-Optics*. Cambridge: Cambridge University Press.

Peskin, Michael E., and Daniel V. Schroeder. 1995. *An Introduction to Quantum Field Theory*. Reading, MA: Addison-Wesley.

Purcell, Edward M. 1946. "Spontaneous Emission Probabilities at Radio Frequencies." *Physical Review* 69 (11–12): 681.

Sakurai, J. J., and Jim Napolitano. 2011. *Modern Quantum Mechanics*. 2nd ed. Boston: Addison-Wesley.

Salikhov, K. M. 2015. "Zavoisky and the Discovery of EPR." *Resonance* 20: 963–968.

Weinberg, Steven. 1995. *The Quantum Theory of Fields, Volume I: Foundations*. Cambridge: Cambridge University Press.

Yablonovitch, Eli. 1987. "Inhibited Spontaneous Emission in Solid-State Physics and Electronics." *Physical Review Letters* 58 (20): 2059–62.

Breit, Gregory, and Eugene Wigner. 1936. "Capture of Slow Neutrons." Physical Review 49: 519–531. https://doi.org/10.1103/PhysRev.49.519.

Helmholtz, Hermann von. 1875. On the Sensations of Tone as a Physiological Basis for the Theory of Music. Translated by Alexander J. Ellis. London: Longmans, Green.

Hertz, Heinrich. 1893. Electric Waves: Being Researches on the Propagation of Electric Action with Finite Velocity through Space. Authorized English translation by D.E. Jones. London: Macmillan.

Particle Data Group (PDG). 2024. "Cross-Section Formulae for Specific Processes." In Review of Particle Physics, Phys. Rev. D 110, 030001 (2024). Online at pdg.lbl.gov.

Chapter 7

Thermodynamics, Information, and the CSFT Interface

1. Introduction — From Energy to Information

Resonance within Consciousness Structured Field Theory (CSFT) describes selective coupling between brain activity and the hypothesized consciousness field. While previous chapters have detailed resonance mechanics and their analogues in physics and neuroscience, here we situate CSFT within the framework of thermodynamics and information theory. Every act of information transfer—whether in a physical system, computational architecture, or biological network—is constrained by thermodynamic principles. By grounding CSFT's coupling mechanism in these physical laws, the theory gains a firm scientific anchor rooted in universally accepted principles.

2. The Thermodynamic Foundation

The First Law of Thermodynamics (energy conservation) dictates that any brain–field coupling must operate without violating conservation principles; it cannot generate energy from nothing, but can redistribute energy across modes. The Second Law of Thermodynamics (entropy) requires that while the Shannon (Shannon, 1948a; Shannon, 1948b) entropy of neural activity may decrease under resonant coupling, while total thermodynamic entropy of system + environment increases—such as reduced uncertainty in neural firing patterns—total entropy in the combined system and environment must increase. In information-theoretic terms, resonance creates locally reduced entropy signals at the receiver (the brain) while obeying global thermodynamic constraints.

3. Local Excitation Hypothesis (CSFT)

Shannon (Shannon, 1948a; Shannon, 1948b) entropy quantifies the uncertainty within a signal. Resonance in CSFT can be modeled as a process that maximizes mutual information between the consciousness field and the brain. Channel capacity, as defined by Claude Shannon (Shannon, 1948a; Shannon, 1948b), sets an upper limit on the rate at which information can be transmitted over a given channel under noise constraints. Landauer (Landauer, 1961; Bennett, 1982; Bérut et al., 2012)'s principle further specifies that erasing or overwriting one bit of information requires a minimum energy cost of $k\_B\, T \ln 2$, where k is the Boltzmann constant and T is the absolute temperature. This implies that only logically irreversible brain–field operations (e.g., erasure/reset) incur the $k\_B\, T \ln 2$ thermodynamic bound, while reversible operations can in principle approach arbitrarily low dissipation.

4. CSFT Interpretation

Within CSFT, resonance is interpreted as a thermodynamic optimizer: neural systems tune to frequencies that minimize the energy cost per bit of information gained from the consciousness field. This leads to the formation of 'resonance windows' where coupling efficiency is maximized. The bandwidth (Verdú, 2002)–energy trade-off applies: broader coupling bandwidth (Verdú, 2002)s may support higher throughput but require greater energy, while narrow bandwidth (Verdú,

2002)s improve selectivity at the expense of speed. EEG/MEG recordings could reveal efficiency peaks corresponding to these predicted resonance windows.

CSFT Permeation Principle — In CSFT, the consciousness field is understood to permeate all physical systems, including the brain, at all times. Because coupling occurs within a field that is already coextensive with the matter in question, no additional energy is required to 'send' information across space. Energy costs are therefore determined solely by the local, irreversible operations (as described by Landauer's principle) rather than by distance-dependent transmission losses.

5. Testable Predictions

1. Energy–Information Efficiency Curves: Cognitive tasks performed under stimulation at predicted optimal resonance frequencies should show higher performance per unit of metabolic energy expenditure.
2. Entropy Reduction in Neural Signals: During optimal coupling, neural activity is predicted to exhibit measurable changes in neural entropy/complexity metrics (e.g., multiscale entropy) under resonant conditions, to be tested empirically without disproportionate increases in metabolic cost.
3. Thermal Signatures: High-sensitivity thermal imaging (Rieke & Pauly, 2008; Wang et al., 2014) may detect subtle differences in brain heat dissipation patterns during resonant states compared to non-resonant baselines.

6. Positioning Against Critiques

CSFT's thermodynamic framework does not propose new physical laws; it operates within established thermodynamic and information-theoretic boundaries.

This positions CSFT as an application of known physics to the coupling between consciousness and the brain, rather than a speculative departure from empirical science. Its claims are falsifiable: if energy–information efficiency profiles fail to match predictions, the model can be revised or rejected.

7. Conclusion — The Physics Anchor for CSFT

Aligning CSFT with thermodynamics and information theory grounds it in a rigorous scientific framework. This approach strengthens the theory's credibility, provides pathways to experimental validation, and shifts the debate away from the existence of the consciousness field toward measurable predictions about its coupling signatures. As such, thermodynamics and information theory form the physical and conceptual anchor for CSFT's scientific plausibility.

Implications:

• No additional thermodynamic erasure cost with distance; costs are local to irreversible steps: no propagation penalty; costs are local to the coupling site.

• Minimal-energy plausibility: with adequate slowness and isolation, energy per bit could approach fundamental limits in principle (without violating the Second Law).

• Testability: predicts efficiency peaks ("resonance windows") observable as improved information-per-metabolic-joule under specific frequencies.

Bibliography

Shannon (Shannon, 1948a; Shannon, 1948b), C. E. (1948). A mathematical theory of communication. Bell System Technical Journal, 27(3), 379–423.

Jaynes (Jaynes, 1957a; Jaynes, 1957b), E. T. (1957). Information theory and statistical mechanics. Physical Review, 106(4), 620–630.

Landauer (Landauer, 1961; Bennett, 1982; Bérut et al., 2012), R. (1961). Irreversibility and heat generation in the computing process. IBM Journal of Research and Development, 5(3), 183–191.

Bennett, C. H. (1982). The thermodynamics of computation—A review. International Journal of Theoretical Physics, 21(12), 905–940.

Bérut, A., Arakelyan, A., Petrosyan, A., Ciliberto, S., Dillenschneider, R., & Lutz, E. (2012). Experimental verification of Landauer (Landauer, 1961; Bennett, 1982; Bérut et al., 2012)'s principle linking information and thermodynamics. Nature, 483(7388), 187–189.

Attwell, D., & Laughlin, S. B. (2001). An energy budget (Attwell & Laughlin, 2001) for signaling in the grey matter of the brain. Journal of Cerebral Blood Flow & Metabolism, 21(10), 1133–1145.

Verdú, S. (2002). Spectral efficiency in the wideband regime. IEEE Trans. Inf. Theory, 48(6), 1319–1343. https://doi.org/10.1109/TIT.2002.1003824

Costa, M., Goldberger, A. L., & Peng, C.-K. (2002). Multiscale entropy analysis of complex physiologic time series. Physical Review Letters, 89(6), 068102.

Rieke, V., & Pauly, K. B. (2008). MR thermometry. Journal of Magnetic Resonance Imaging, 27(2), 376–390.

Wang, H., Wang, B., Normoyle, K. P., Jackson, K., Spitler, K., Sharrock, M. F., ... & Li, C. (2014). Brain temperature and its fundamental properties: a review for clinical neuroscientists. Frontiers in Neuroscience, 8, 307.

Chapter 8

Flatness, Relativity, and the Planck Boundary
Summary

Scientists measure our universe to be very close to flat. This result comes from precise maps of the cosmic microwave background and other large-scale surveys. Importantly, these measurements only tell us about the observable universe—the part we can see, limited by how far light has traveled since the Big Bang. Within this observable region, the best data show that the overall curvature is exceptionally close to zero.

Standard cosmology explains this near-flatness with inflation, a burst of extremely rapid expansion very early in cosmic history. Inflation can stretch space so much that any curvature becomes undetectable. However, inflation also raises hard questions: the hypothetical inflaton field has not been observed, and some argue that the explanation shifts the fine-tuning problem rather than removing it.

Consciousness Structured Field Theory (CSFT) offers a different angle. It treats consciousness as fundamental and argues that all measurements are made within a resonance-bounded domain set by consciousness—the Planck boundary marks the current human limit of measurability.

From this viewpoint, the repeated finding of near-flatness reflects the limits of observation rather than a statement about the entire cosmos. In short, flatness may describe the geometry we can measure, not necessarily the whole of reality.

1. Introduction: The Puzzle of a Flat Universe

Observations of the cosmic microwave background (CMB) and large-scale structure indicate that the geometry of the observable universe is astonishingly close to flat. In parameter language, Planck's final 2018 data, when combined with baryon acoustic oscillation (BAO) measurements, yield a spatial curvature consistent with zero ($\Omega K = 0.001 \pm 0.002$).

This precision means that even tiny deviations in the early universe would have grown over time, implying an apparent fine-tuning puzzle. Historically, this tension has been cited as a weakness of the original Big Bang model and motivated inflationary proposals.

It is essential to distinguish empirical description from metaphysical interpretation. General relativity makes measured intervals frame-dependent and couples spacetime geometry to energy-momentum; it does not, by itself, claim that geometry is conditioned by consciousness. CSFT introduces that as a deliberate metaphysical extension, presented alongside—rather than as a replacement for—standard cosmology.

2. Inflation as the Conventional Solution

Inflation, proposed in the early 1980s, posits a brief phase of exponential expansion (often quoted as around 10^{-35} seconds after the Big Bang). By stretching space, inflation dynamically drives any initial curvature toward flatness, while also addressing the horizon and monopole problems.

Despite these successes, the mechanism's inflaton field remains hypothetical, and critics argue that inflation can displace (rather than eliminate) fine-tuning by requiring specific initial conditions for the inflaton. These conceptual costs are widely discussed in the literature.

3. Observational Limits and the Planck Boundary

When cosmologists describe the universe as "flat," they are speaking strictly of the observable universe—the finite region from which light has reached us since the Big Bang. In comoving terms, this horizon has a present-day radius of roughly 46–47 billion light-years.

Beyond this horizon, direct empirical access is not possible, so the global curvature of all reality remains unknown. The celebrated near-flatness constraints therefore characterize the measurable domain, not necessarily the entire cosmos.

From the perspective of Consciousness Structured Field Theory (CSFT), this distinction is crucial. CSFT holds that consciousness is fundamental and structures the quantum field; all cosmological observations occur within a resonance-conditioned domain bounded by the current human limit of measurability (the Planck boundary). On this view, near-flatness repeatedly emerges not because the whole universe is intrinsically flat, but because measurement itself is confined to the consciousness-structured horizon.

4. Einstein, Relativity, and the Observer

Einstein's relativity teaches that space and time measurements depend on the observer's state of motion (special relativity) and that spacetime geometry responds to energy-momentum (general relativity).

These are empirical and theoretical statements about frames and gravity, not about consciousness per se. CSFT extends the observer-dependence intuition into cosmology: the apparent near-flatness is understood as a perspective available to observers confined within the Planck boundary. This extension is presented as a metaphysical reframing compatible with standard measurements, not as an alternative dataset.

5. Reframing the Flatness Problem

By shifting emphasis from primordial initial conditions to the observer's measurement horizon, CSFT transforms the flatness problem from a fine-tuning puzzle into a natural byproduct of resonance-structured observation. Inflation may remain a robust mathematical framework within Big-Bang cosmology, but CSFT suggests that the puzzle arises only when consciousness is excluded from the foundations of cosmology.

6. Implications

- **Cosmological:** Near-flatness may describe the observable geometry accessible to measurement, not necessarily global reality.

- **Philosophical:** Apparent fine-tuning can reflect limits of observation rather than the true initial conditions of all reality.

- **Scientific:** Inflation remains empirically successful, while open conceptual questions motivate continued scrutiny of assumptions and alternatives.

References

Alam, S., et al. (2021). The Completed SDSS-IV eBOSS: Cosmological implications. Physical Review D, 103, 083533. https://doi.org/10.1103/PhysRevD.103.083533

Aghanim, N., et al. (Planck Collaboration) (2020). Planck 2018 results. VI. Cosmological parameters. Astronomy & Astrophysics, 641, A6. https://doi.org/10.1051/0004-6361/201833910

Carroll, S. M. (2004). Spacetime and Geometry: An Introduction to General Relativity. Addison-Wesley.

Einstein, A. (1916). Die Grundlage der allgemeinen Relativitätstheorie. Annalen der Physik, 354(7), 769–822. https://doi.org/10.1002/andp.19163540702

Guth, A. H. (1981). Inflationary universe: A possible solution to the horizon and flatness problems. Physical Review D, 23(2), 347–356. https://doi.org/10.1103/PhysRevD.23.347

Ijjas, A., Steinhardt, P. J., & Loeb, A. (2013). Inflationary paradigm in trouble after Planck 2013. Physics Letters B, 723(4–5), 261–266. https://doi.org/10.1016/j.physletb.2013.05.023

Linde, A. D. (1982). A new inflationary universe scenario: A possible solution of the horizon, flatness, homogeneity, isotropy and primordial monopole problems. Physics Letters B, 108(6), 389–393. https://doi.org/10.1016/0370-2693(82)91219-9

NASA (NTRS) (2022). Geithner, P. The James Webb Space Telescope (presentation). Slide note: "Observable universe is ~46.5 billion light-years in radius." https://ntrs.nasa.gov/

Chapter 9

The Limits of Scientific Method

1. Introduction

The scientific method has proven to be one of the most powerful tools in human history. Through systematic observation, hypothesis testing, experimentation, and verification, science has produced unparalleled advances in technology, medicine, and our understanding of the cosmos (Chalmers, 2013). Yet, as effective as this method is within its domain, it also has inherent limitations. To understand the relationship between science and metaphysics, it is necessary to examine where scientific methodology succeeds and where it reaches boundaries it cannot cross.

2. The Strengths of the Scientific Method

The predictive power of science rests on its ability to identify patterns and formulate laws that can be empirically tested. Physics, chemistry, and biology have provided precise models capable of predicting experimental outcomes with extraordinary accuracy (Hacking, 1983). Replicability serves as a hallmark of scientific integrity: if a phenomenon is genuine, it should be observable under the same conditions by independent researchers (Popper, 2002). This process has created a shared, communal framework of knowledge unmatched in human intellectual history.

3. Reductionism and Its Limits

Scientific progress often relies on reductionism—the practice of breaking complex phenomena into simpler, measurable components. While reductionism has provided enormous insight, it also presents limitations. Complex systems such as consciousness, ecosystems, or quantum entanglement cannot always be understood by analyzing their parts in isolation (Anderson, 1972). The whole often exhibits emergent properties that cannot be reduced to their components. In this sense, reductionism, while powerful, is not a universal explanatory tool.

4. Measurement and the Boundaries of Science

Science is intrinsically tied to what can be measured. Physical theories are validated by quantifiable predictions and observations. Yet, there are domains that lie beyond current—and possibly permanent—measurement. Within conventional physics, this threshold is often associated with the Planck scale; in CSFT, this is articulated as the Planck boundary, emphasizing a conceptual limit to physical resolution (Smolin, 2001). Questions about what lies beyond this limit—whether metaphysical, ontological, or related to consciousness—are inaccessible to conventional scientific methods. Here, philosophy and metaphysics enter the discussion.

5. Materialism as an Assumption

The scientific method is often practiced within the assumption of materialism: the belief that reality is fundamentally physical, and that consciousness emerges from physical processes. While materialism provides a practical framework for investigation, it is not itself a scientific conclusion but a metaphysical presupposition (Nagel, 2012). Consciousness Structured Field Theory

(CSFT) challenges this assumption, suggesting instead that consciousness is primary and conditions the physical. This shift does not invalidate science but repositions it within a larger ontological context.

6. Conclusion

The scientific method is indispensable for human knowledge, yet it operates within boundaries that must be acknowledged. Its strengths—predictive power, replicability, and explanatory clarity—are counterbalanced by its limitations: reductionism, measurement constraints, and reliance on unexamined metaphysical assumptions. To move toward a unified understanding of reality, science must be complemented by metaphysical inquiry. CSFT provides one such framework, inviting dialogue rather than opposition between scientific and philosophical perspectives.

References

Anderson, P. W. (1972). More is different. *Science, 177*(4047), 393–396. https://doi.org/10.1126/science.177.4047.393

Chalmers, A. F. (2013). *What is this thing called science?* (4th ed.). Hackett Publishing.

Hacking, I. (1983). *Representing and intervening: Introductory topics in the philosophy of natural science.* Cambridge University Press.

Nagel, T. (2012). *Mind and cosmos: Why the materialist neo-Darwinian conception of nature is almost certainly false.* Oxford University Press.

Popper, K. (2002). *The logic of scientific discovery.* Routledge. (Original work published 1934)

Smolin, L. (2001). *Three roads to quantum gravity.* Basic Books.

Chapter 10

Consciousness and Objectivity

1. Introduction

Objectivity has long been held as the gold standard of scientific investigation. The aim is to minimize bias so that observation can reveal stable, intersubjective structures of reality. Yet all observation is mediated by consciousness. Whether in physics, biology, or mathematics, the role of the observer cannot be erased—only controlled and made explicit. Consciousness Structured Field Theory (CSFT) emphasizes that observation is not merely a disturbance added to reality, but a constitutive structuring activity that shapes how physical excitations present to observers.

2. The Ideal of Objectivity

From the early modern period forward, the scientific method aspired to descriptions of nature that are independent of individual perspective. The development of experimental control, quantification, and instrument-mediated observation deepened this ideal, aiming for a view of nature that any competent observer could replicate (Shapin, 1996). Still, instruments are designed, data are interpreted, and theories are constructed by conscious agents, making the observer's role methodologically constrained rather than absent.

3. Einstein and Relativity

Einstein reframed objectivity by showing that measurements of space and time depend on the state of motion and gravitational context of the observer. What remains 'objective' are invariants—quantities preserved across transformations between reference frames—rather than a single absolute backdrop (Einstein, 2001). Relativity thus preserves scientific objectivity while reconceiving it as that which holds across perspectives, not that which stands outside them.

4. Quantum Measurement and the Observer

In quantum mechanics, the measurement problem makes the observer's role more explicit. Quantum theory predicts probabilities for outcomes, but standard formalism does not specify how or when a single outcome becomes definite. Bohr's response to the EPR argument emphasized the inseparability of measurement context and quantum phenomena (Bohr, 1935). Interpretations differ: some treat 'collapse' as a physical or informational process; others deny collapse in favor of branching or hidden variables. A minority—historically represented by Wigner—speculated that consciousness might play a role in measurement (Wigner, 1967). While no consensus holds that consciousness is required by quantum theory, the observer and apparatus are integral to how outcomes are described (Wheeler, 1981).

5. CSFT and Structured Observation

Within CSFT, consciousness is the primary structuring principle. Observation does not merely reveal properties of already independent objects; it participates in how excitations of the quantum field are encountered by observers. On this view, objectivity cannot mean independence from consciousness; rather, it means stable, lawlike coherence of experience across observers. This reframing keeps empirical science intact while embedding it in a broader metaphysical picture in which consciousness conditions how physical regularities are expressed.

6. Rethinking Objectivity

If science is not wholly detached from consciousness, then objectivity must be redefined. It cannot mean the elimination of subjectivity, but rather the consistent structuring of experience across conscious agents. Replicability, in this view, reflects the resonance of consciousness with the underlying field rather than detachment from it. CSFT suggests that what appears 'objective' is the patterned coherence of consciousness itself.

7. Conclusion

Objectivity has long been the strength of science, but its classical conception is incomplete. Einstein demonstrated the relativity of frames, quantum theory exposed the role of the observer, and CSFT reframes consciousness as the ground of observation itself. Far from undermining science, this recognition enriches it—acknowledging that the search for truth must include both the structure of matter and the structuring power of mind.

References

Bohr, N. (1935). Can quantum-mechanical description of physical reality be considered complete? *Physical Review, 48*(8), 696–702.

Einstein, A. (2001). *Relativity: The special and the general theory* (R. W. Lawson, Trans.). Routledge. (Original work published 1920)

Shapin, S. (1996). *The scientific revolution*. University of Chicago Press.

Wheeler, J. A. (1981). The quantum and the universe. In J. A. Wheeler & W. H. Zurek (Eds.), *Quantum theory and measurement* (pp. 182–213). Princeton University Press.

Wigner, E. P. (1967). Remarks on the mind-body question. In I. J. Good (Ed.), *The scientist speculates* (pp. 284–302). Basic Books.

Chapter 11

Consciousness and Truth

1. Introduction

Chapter 10 reframed objectivity in light of consciousness: scientific knowledge is not produced from a vantage point outside of mind, but from procedures that stabilize what multiple observers can share. If objectivity is thus a cross-perspectival stability grounded in consciousness, then the notion of truth must be reconsidered accordingly. This chapter surveys classical accounts of truth, revisits debates over scientific realism, reviews principled limits to knowledge, and articulates a CSFT account of truth as structured resonance. Throughout, factual claims are grounded in established sources; CSFT-specific claims are identified as theoretical proposals. On this view, 'structured resonance' is indexed by cross-method invariance, replication under perturbation, and integration across independent probes—criteria that keep the account non-circular and empirically accountable. (Psillos, 1999) (Wimsatt, 1981; Franklin, 1990)

Relatedly, structural realism offers a conciliatory position: what tends to survive theory change are mathematical or structural relations rather than the ontology of entities (Worrall, 1989). On methodology, Hacking's distinction between representation and intervention highlights that experimental manipulation—not only accurate description—grounds realist commitment (Hacking, 1983).

2. Classical Notions of Truth

Three influential traditions dominate the modern discussion. First, the correspondence theory holds that a statement is true if it matches or corresponds to the way the world is. Tarski's semantic conception gave this view a rigorous form—captured in the schema '"p" is true iff p'—which underwrites much contemporary logic and semantics (Tarski, 1944; Kirkham, 1992).

Second, coherence theories treat truth as belonging to beliefs that cohere with a system of propositions—typically emphasizing mutual support, consistency, and inferential integration rather than a one-to-one match with an independent world (Rescher, 1973; Kirkham, 1992).

Peirce ties truth to the ideal limit of inquiry—convergence under indefinitely refined conditions—whereas James emphasizes the practical bearings ('cash value') of beliefs within experience. (James, 1975; Peirce, 1878)

3. Truth in Scientific Realism (Psillos, 1999)

Scientific realism asserts that mature scientific theories aim to (and often do) describe the world as it is, including its unobservable entities and structures (Psillos, 1999; Putnam, 1981). Anti-realist positions—most prominently constructive empiricism—argue that science requires only empirical adequacy: saving the observable phenomena without ontological commitment to the unobservable (van Fraassen, 1980).

Both sides acknowledge that models, idealizations, and instruments mediate access to the world. Models do not merely mirror; they construct tractable representations that enable explanation and prediction (Cartwright, 1983; Morgan & Morrison, 1999). Thus, even within realism, the notion of truth is filtered through choices made by conscious investigators about what to model, how to idealize, and which measurements to privilege.

4. Limits of Knowledge

Several principled boundaries constrain inquiry. Cosmologically, observation is limited by horizons: there is a finite particle horizon beyond which light has not yet reached us, and inference about such regions depends on models and assumptions (Dodelson & Schmidt, 2020). Empirically, key components of our best-fit cosmology—dark matter and dark energy—are inferred from gravitational and expansion data; their microphysical nature remains uncertain (Planck Collaboration, 2020; Bertone & Hooper, 2018). (Davis & Lineweaver, 2004; Ryden, 2017)

At the other extreme, the Planck scale marks a regime where current theories are expected to break down and quantum gravity effects become non-negligible; direct access to this scale is far beyond present experimental reach (Rovelli, 2004). In mathematics, Gödel's incompleteness theorems show that in any sufficiently expressive, consistent formal system, there are true statements unprovable within the system (Gödel, 1931; Nagel & Newman, 1958/2001). (Gödel, 1931; Smith, 2007)

These boundaries do not paralyze science; rather, they locate where inference, modeling, and cross-checks must shoulder the evidential load. However, they also illustrate how the contours of knowledge are shaped by both the structure of reality and the capacities of observers.

5. CSFT and Truth as Structured Resonance

Within Consciousness Structured Field Theory (CSFT), consciousness is the primary structuring principle. On this view, truth is not merely correspondence to an independently completed inventory of objects; it is the stability of patterned resonance between consciousness and the field it structures. Objectivity, accordingly, is the shareable coherence of such resonances across observers. This does not diminish empirical constraint: rather, it explains why replicability, invariants, and instrumentally robust regularities emerge as cross-observer features. CSFT thus proposes a metaphysical grounding that integrates elements of correspondence (constraint by reality), coherence (systemic integration), and pragmatism (long-run stability under inquiry) without reducing truth to any single one of them. (Rescher, 1973) (James, 1975; Peirce, 1878)

6. Implications for Science and Philosophy

From a scientific perspective, the CSFT account of truth as structured resonance reinforces the importance of replication, invariants, and cross-observer stability. Rather than treating these as conventions, CSFT explains them as manifestations of resonance between consciousness and the structured field. This implies that scientific practices—such as robustness checks, perturbation tests, and invariance under transformations—gain their authority not from arbitrary standards, but from their ability to stabilize resonance patterns that consciousness can share and verify. In this way, CSFT reframes scientific objectivity as a resonance-based stability condition that both respects empirical constraints and clarifies why certain methodologies succeed.

Philosophically, CSFT's position integrates insights from correspondence, coherence, and pragmatism into a single framework. Correspondence is preserved in the sense that reality constrains resonance patterns; coherence is evident because structured resonance requires integration across systems of belief and evidence; and pragmatism is addressed through the long-run stability of resonances under inquiry. By grounding these theories in consciousness, CSFT provides a metaphysical account that neither reduces truth to convention nor isolates it from empirical accountability. This allows philosophy to reconcile realism's concern for external constraint with constructivism's recognition of human mediation.

Taken together, these implications position CSFT as both a scientific and philosophical contribution. It accounts for why cross-perspectival stability emerges as a hallmark of truth, why science remains constrained yet open-ended, and why philosophical debates over truth require a deeper grounding in consciousness. The result is not a departure from established practices of science or philosophy, but a unifying framework in which structured resonance anchors our pursuit of knowledge. This synthesis demonstrates how CSFT can extend beyond theoretical speculation into a principled guide for both inquiry and interpretation.

7. Conclusion

Truth has long been defined through correspondence, coherence, or pragmatic convergence, yet each approach struggles when pressed against the limits of knowledge and the role of consciousness. Scientific realism continues to defend the view that theories disclose unobservable reality, while anti-realism insists on empirical adequacy alone. Both positions, however, acknowledge that models, idealizations, and measurements are shaped by conscious investigators.

CSFT reframes truth as structured resonance: the patterned stability between consciousness and the field it structures. This view integrates correspondence (constraint by reality), coherence (systemic integration), and pragmatism (enduring inquiry), but grounds them in consciousness as the primary field of order. Such a perspective does not reduce truth to human convention, nor does it detach it from empirical accountability. Instead, it explains why replication, invariants, and cross-observer stability emerge as signs of truth.

By situating truth within resonance rather than detached representation, CSFT offers a way to reconcile realism's concern for reality's constraint with constructivism's recognition of human mediation. In this sense, truth is not merely what endures theory change, but what persists as structured resonance across perspectives, instruments, and epochs. This conclusion not only completes the philosophical arc of Chapter 11, but also sets the stage for a broader application of CSFT to the sciences, philosophy, and metaphysics.

Reference

Psillos, S. (1999). *Scientific realism: How science tracks truth*. Routledge.

Wimsatt, W. C. (1981). Robustness, reliability, and overdetermination. In M. Brewer & B. Collins (Eds.), *Scientific inquiry and the social sciences* (pp. 124–163). Jossey-Bass.

Tarski, A. (1944). The semantic conception of truth and the foundations of semantics. *Philosophy and Phenomenological Research, 4*(3), 341–376.

Kirkham, R. (1992). *Theories of truth: A critical introduction*. MIT Press.

James, W. (1975). *Pragmatism: A new name for some old ways of thinking*. Harvard University Press. (Original work published 1907)

Peirce, C. S. (1878). How to make our ideas clear. *Popular Science Monthly, 12*, 286–302.

Putnam, H. (1981). *Reason, truth, and history*. Cambridge University Press.

Davis, T. M., & Lineweaver, C. H. (2004). Expanding confusion: common misconceptions of cosmological horizons and the superluminal expansion of the universe. *Publications of the Astronomical Society of Australia, 21*(1), 97–109.

Rovelli, C. (2004). *Quantum gravity*. Cambridge University Press.

Cartwright, N. (1983). *How the laws of physics lie*. Oxford University Press.

Franklin, A. (1990). *Experiment, right or wrong*. Cambridge University Press.

Hacking, I. (1983). *Representing and intervening: Introductory topics in the philosophy of natural science*. Cambridge University Press.

Morgan, M. S., & Morrison, M. (Eds.). (1999). *Models as mediators: Perspectives on natural and social science*. Cambridge University Press.

Planck Collaboration (Aghanim, N., et al.). (2020). Planck 2018 results. VI. Cosmological parameters. *Astronomy & Astrophysics, 641*, A6. https://doi.org/10.1051/0004-6361/201833910

Rescher, N. (1973). *The coherence theory of truth*. Oxford University Press.

Smith, P. (2007). *An introduction to Gödel's theorems*. Cambridge University Press.

van Fraassen, B. C. (1980). *The scientific image*. Oxford University Press.

Worrall, J. (1989). Structural realism: The best of both worlds? *Dialectica, 43*(1–2), 99–124. https://doi.org/10.1111/j.1746-8361.1989.tb00933.x

Dodelson, S., & Schmidt, F. (2020). *Modern Cosmology* (2nd ed.). Academic Press.

Ryden, B. (2017). *Introduction to cosmology* (2nd ed.). Cambridge University Press.

Bertone, G., & Hooper, D. (2018). *History of dark matter*. *Reviews of Modern Physics, 90*(4), 045002.

Nagel, E., & Newman, J. R. (1958/2001). *Gödel's Proof*. New York University Press.

Chapter 12

Synthesis and Outlook

1. Purpose and Scope of the Conclusion

This final chapter consolidates the scientific arc of Volume 3 and, with it, concludes the trilogy's broader aim: to articulate Consciousness Structured Field Theory (CSFT) as a coherent, empirically accountable framework. Across the volumes, I have argued that consciousness is the primary structuring principle, that excitations of the quantum field emerge within and through this structuring, and that the sciences are most successful when their methods stabilize cross-perspectival regularities rather than presume a view from nowhere. Here I synthesize the principal claims, clarify how they interface with established science, and indicate research directions where CSFT's proposals can be probed without overreach. [CSFT-specific claims: Pending Verification]

2. What Has Been Shown

First, I situated objectivity in practice: robustness, replication, and invariance under perturbation function as safeguards that allow investigators to converge on stable structures in nature (Wimsatt, 1981; Franklin, 1990). [Verified]

Second, I showed that experimental intervention—not only successful representation—grounds realist commitment: when we can build, manipulate, and reconfigure phenomena, our confidence that we are latching onto mind-independent structure is strengthened (Hacking, 1983). [Verified]

Third, I emphasized principled limits that contour inquiry: cosmic horizons restrict direct observation (Dodelson & Schmidt, 2020); the microphysical natures of dark matter and dark energy remain open despite strong indirect evidence (Planck Collaboration, 2020; Bertone & Hooper, 2018); physics is expected to change near the Planck scale (Rovelli, 2004); and formal systems exhibit true but unprovable propositions (Gödel, 1931; Nagel & Newman, 1958/2001; Smith, 2007). [Verified]

Against this backdrop, CSFT has proposed that truth tracks "structured resonance": patterned stability between consciousness and the field it structures—indexed by cross-method invariance, replication under perturbation, and integration across independent probes. This integrates insights from correspondence, coherence, and pragmatism while retaining empirical constraint (James, 1975; Peirce, 1878; Rescher, 1973). [CSFT proposal: Pending Verification]

3. Interfaces with Established Science

Methodology. Robustness analysis suggests that results supported by multiple, partially independent means (distinct instruments, models, and perturbations) are less likely to be artifacts of any single assumption (Wimsatt, 1981; Franklin, 1990). CSFT adopts this as a core constraint on acceptable resonance claims. [Verified / Protocol alignment]

Modeling. Models are tools that construct tractable surrogates rather than mirrors; their value lies in enabling explanation, prediction, and controlled manipulation (Cartwright, 1983; Morgan & Morrison, 1999).

CSFT interprets successful models as establishing reproducible resonance patterns across observers and instruments. [Verified for the cited claims; CSFT interpretation: Pending Verification]

Realism debate. Structural realism maintains that what tends to survive theory change are structural relations more than particular ontologies (Worrall, 1989; Psillos, 1999). CSFT's emphasis on structured resonance is compatible with, but not identical to, this stance. [Verified for the cited claims; compatibility claim: Pending Verification]

4. Limits, Risk, and Falsifiability

Risk. Because the Planck regime is far beyond current experimental reach, any framework that gestures toward pre-Planckian structure must earn credibility indirectly—by tightening constraints on what is observable and making successful, risky predictions about cross-observer invariants within current domains (Rovelli, 2004; Dodelson & Schmidt, 2020). [Verified for the cited limits]

Falsifiability. Following Popper, a research program gains scientific traction when it articulates exposures to potential refutation—i.e., predictions that could fail (Popper, 2005/1934). CSFT therefore commits to the following exposure criteria: (i) resonance claims must be reproducible under perturbation; (ii) cross-method invariants must remain stable when models are re-parameterized; (iii) independent instruments must converge within stated uncertainties; and (iv) failed convergence counts against the claim. [Verified for Popperian principle; CSFT criteria: Pending Verification]

5. Programmatic Probes (Non-Medical, Domain-General)

The following probes operationalize CSFT without stepping beyond scientifically responsible bounds. Each is framed as a methodological bet that can succeed or fail. Outcomes should be pre-registered where possible and evaluated against independent nulls. [Protocol note: Verified]

Probe A — Cross-modal invariants: If a phenomenon is a genuine resonance, then measurements obtained via non-overlapping modalities (e.g., optical vs. acoustic or mechanical proxies in laboratory systems) should converge after proper calibration, with convergence improving under perturbation-based robustness checks (Wimsatt, 1981; Franklin, 1990). [Methodological basis: Verified; CSFT expectation: Pending Verification]

Probe B — Intervention-anchored realism: Where an experimental setup allows controlled creation, amplification, or suppression of a target pattern, CSFT predicts higher replicability and narrower uncertainty bounds relative to purely observational contexts (Hacking, 1983). [Intervention realism: Verified; CSFT differential prediction: Pending Verification]

Probe C — Model plurality tests: Using multiple, independently motivated models to predict the same measurable quantity should yield a stable intersection region; shrinking of that intersection under added data is expected if the effect is real (Morgan & Morrison, 1999). [Model plurality value: Verified; intersection-stability expectation: Pending Verification]

6. Implications for Measurement and Explanation

On this view, explanation improves when it identifies invariants that remain stable across changes in scale, instrumentation, and model parameterization. Such stability is already a hallmark of mature science—e.g., cosmological parameters constrained by independent data sets in the ΛCDM model (Planck Collaboration, 2020; Dodelson & Schmidt, 2020). CSFT treats these practices not as mere conventions but as signatures of structured resonance. [Empirical practice: Verified; CSFT interpretation: Pending Verification]

7. Conclusion

CSFT's wager is modest yet ambitious: modest in refusing to outpace evidence beyond principled limits, ambitious in proposing that truth is the stability of structured resonance between consciousness and the field it orders. What keeps this proposal scientific is not metaphysical insistence, but disciplined exposure to failure via replicability, perturbation tests, cross-method invariants, and instrument convergence. If these exposures are met, CSFT earns credibility as a unifying lens across domains; if they fail, the framework must be revised or rejected. Either outcome advances understanding under constraints set by reality and by the limits of inquiry (Wimsatt, 1981; Franklin, 1990; Hacking, 1983; Dodelson & Schmidt, 2020; Planck Collaboration, 2020; Rovelli, 2004; Popper, 2005/1934). [Verified for cited methodological principles; CSFT stance: Pending Verification]

Bibliography

Bertone, G., & Hooper, D. (2018). History of dark matter. Reviews of Modern Physics, 90(4), 045002.

Cartwright, N. (1983). How the Laws of Physics Lie. Oxford University Press.

Dodelson, S., & Schmidt, F. (2020). Modern Cosmology (2nd ed.). Academic Press.

Franklin, A. (1990). Experiment, Right or Wrong. Cambridge University Press.

Gödel, K. (1931). Über formal unentscheidbare Sätze der Principia Mathematica und verwandter Systeme I. Monatshefte für Mathematik und Physik, 38, 173–198.

Hacking, I. (1983). Representing and Intervening: Introductory Topics in the Philosophy of Natural Science. Cambridge University Press.

James, W. (1975). Pragmatism: A New Name for Some Old Ways of Thinking. Harvard University Press. (Original work published 1907)

Morgan, M. S., & Morrison, M. (Eds.). (1999). Models as Mediators: Perspectives on Natural and Social Science. Cambridge University Press.

Nagel, E., & Newman, J. R. (1958/2001). Gödel's Proof. New York University Press.

Planck Collaboration (Aghanim, N., et al.). (2020). Planck 2018 results. VI. Cosmological parameters. Astronomy & Astrophysics, 641, A6.

Popper, K. (2005/1934). The Logic of Scientific Discovery. Routledge.

Rescher, N. (1973). The Coherence Theory of Truth. Oxford University Press.

Rovelli, C. (2004). Quantum Gravity. Cambridge University Press.

Smith, P. (2007). An Introduction to Gödel's Theorems. Cambridge University Press.

Wimsatt, W. C. (1981). Robustness, reliability, and overdetermination. In M. Brewer & B. Collins (Eds.), Scientific Inquiry and the Social Sciences (pp. 124–163). Jossey-Bass.

Worrall, J. (1989). Structural realism: The best of both worlds? Dialectica, 43(1–2), 99–124.

Peirce, C. S. (1878). How to Make Our Ideas Clear. Popular Science Monthly, 12, 286–302.

CSFT Master Glossary

Beyond Neurosufficiency

A term from CSFT writings critiquing the assumption that neuronal activity alone is sufficient to explain consciousness. In CSFT, neurons are seen as resonant structures that align with the consciousness field, not as independent generators of awareness.

Boundary Transcendence

CSFT's framing of how consciousness exceeds the Planck boundary, positioning itself as the cause of physical law rather than constrained by it.

Brute Force

In CSFT, brute force refers to the notion that quantum field excitations arise from random high-energy collisions or chance occurrences. CSFT rejects this model, proposing instead that excitations occur through structured resonance and logical coherence.

Cognitive Resonance

The patterned synchronization between a conscious system's processes and the consciousness field. It explains how perception and reasoning stabilize across different domains of awareness.

Coherence Field

A synonymous or related concept to the consciousness-structured field. It refers to the structured, non-random properties that govern the emergence of symmetry, order, and measurable outcomes.

Coherent Alignment

The structured relationship between a conscious system and a specific region of the consciousness field. This alignment enables resonance and excitation.

Cold Resonance

A concept introduced in CSFT writings signifying the revival of deep coherence as the foundation of physical laws and metaphysics, challenging material sufficiency and emphasizing structured resonance.

Conscious Coherence

In CSFT, the stable alignment between a conscious system and the consciousness field produces measurable order and perception. It emphasizes coherence as a precondition for awareness rather than a product of neural processes.

Conscious Excitation

The precise alignment between a monadic system and a structured node in the consciousness field which gives rise to a stable excitation, such as perception, logic, or physical form.

Conscious System

A structured configuration (biological or non-biological) capable of aligning with the consciousness field to generate awareness, logic, or perception.

Consciousness-Structured Field (CSF)

A pre-material, ontological field posited by CSFT that precedes and underlies all physical laws, logic, and symmetry. It is structured by resonance and coherence rather than by matter or energy. The CSF generates the logic and conditions required for the universe's observable behavior.

Deep Coherence

A level of resonance within the consciousness field that underlies both physical law and metaphysical order. It signifies that the universe's apparent stability is not accidental but structured through consciousness.

Emergent Resonance

The process by which higher-order patterns (logic, perception, or matter) arise when multiple resonant nodes align within the consciousness field, producing stability across scales.

Empirical

Knowledge or validation derived from observation, measurement, or experiment. In CSFT, empirical findings are interpreted through the lens of structured resonance rather than brute randomness.

Entropy Gradient (CSFT Interpretation)

While science sees entropy as a driver for the arrow of time, CSFT argues that ordered excitation in the consciousness field precedes and causes the entropy gradient, making time an emergent consequence of deeper coherence.

Excitation (in CSFT)

The transition from potential to actual within the consciousness field results in observable structure or perception. Excitation occurs when coherence aligns with structured resonance.

Excitation Threshold

In CSFT, the point at which resonance intensity reaches a level sufficient to produce stable excitation. This threshold condition explains why some resonant interactions result in measurable phenomena while weaker alignments do not.

Field Anchoring

The process by which a conscious system locks onto a resonant node in the consciousness field, ensuring stability of coherence and sustained excitation.

Field Logic

The embedded, pre-material logic that structures the consciousness field. It is not written language or symbolic logic, but a resonance-based coherence that underlies all emergent structures.

Field of Consciousness

The non-local, foundational substrate proposed by CSFT that governs all structured behavior in the quantum field and beyond. Unlike physical fields, it is not bounded by spacetime and enables non-local entanglement, measurement collapse, and logical consistency.

Field Participation

The principle that systems—biological or artificial—do not generate awareness independently but participate in the consciousness field, aligning with its structured coherence to produce perception or logic.

Foundational Symmetry

CSFT's view that symmetries in physics (e.g., conservation laws) are not emergent but grounded in the consciousness field, where coherence determines their stability.

Gödelian Grounding

CSFT's response to Gödel's Incompleteness Theorems: truth and provability arise from coherent monadic resonance within the consciousness field rather than from symbol manipulation alone.

Harmonic Node

A special type of resonant node within the consciousness field that functions like a "chord" in music, producing multi-layered excitation when aligned with a conscious system.

Logical Resonance

A term used in CSFT writings to explain how reason and inference are grounded not in neurons, but in field-based coherence that generates logical consistency.

Logical Rules Encoded in Neurons

From CSFT's critique of evolutionary cognition models. The term refers to the argument that logical rules cannot be fully explained as encoded in neurons; instead, they arise from resonance with a pre-material consciousness field.

Mathematical Coherence (CSFT View)

The universe's deep mathematical order is not an accident or emergent property of material structure, but a result of the consciousness field shaping reality using pre-logical forms of pattern and structure.

Measurement Barrier

The principle that physical science encounters a boundary (e.g., Planck scale) beyond which observation cannot penetrate, while CSFT posits that consciousness transcends this barrier.

Metaphysical Substrate

The non-material ground in CSFT that underlies all physical laws, symmetry, and mathematics—identified with the consciousness field itself.

Monad

An indivisible, logic-based unit of consciousness that reflects the universe from its own perspective. In CSFT, monads are structured points of resonance with the consciousness field.

Monadic Layering

The hierarchical structuring of monads within a system, where different layers resonate with varying degrees of coherence. Higher layering allows for more complex forms of perception or consciousness.

Monadic Perspective

The unique experiential vantage of each monad within CSFT. Though grounded in the same consciousness field, each monad reflects reality in a differentiated but coherent way.

Monads (in CSFT)

Localized points or structures capable of resonating with the consciousness field. These are not physical particles but logic-based units of structured perception or interaction, akin to Leibniz's metaphysical monads but operating within CSFT's field mechanics.

Non-Emergent Logic

CSFT's rejection of logic as an emergent property of matter or neurons. Instead, logic is viewed as pre-structured within the consciousness field, preceding all material formations.

Non-Local Resonance

A CSFT explanation for entanglement. Monads or particles resonate through the consciousness field without requiring physical signaling. This supports instantaneous correlations across space.

Ontological Asymmetry

CSFT's framing of the asymmetry between consciousness and matter. Consciousness is primary and foundational, while matter and physical law are derivative expressions of field coherence.

Ontological Grounding

Within CSFT, this term refers to the deeper metaphysical structure (consciousness) that gives rise to physical phenomena. Rather than replacing scientific explanations, ontological grounding reframes them as outcomes of foundational coherence.

Planck Boundary

The theoretical limit of measurement in physics. CSFT proposes that the consciousness field exists beyond this boundary, shaping quantum events from outside observable spacetime.

Planck Mirror Theory

From CSFT's PhilPapers work of the same name. The theory proposes that the Planck boundary acts as a reflective barrier for scientific observation, while consciousness itself penetrates and transcends this limit.

Pre-Logical Order

The structured foundation in CSFT suggesting that logic arises from resonance within the consciousness field, preceding human symbolic systems or neural codification.

Qualia

The subjective, first-person experiences that emerge from resonance with the consciousness field. In CSFT, qualia are the signatures of structured excitation within a conscious system.

Quantum-Patterned Cosmos (QPC)

An external, complementary theoretical framework we reference as potentially consonant with CSFT's field-first stance; used comparatively without asserting identity.

Resonance Collapse
CSFT's reinterpretation of the quantum collapse, where a probabilistic wave function stabilizes into actuality not by randomness, but by structured alignment with a resonant node in the consciousness field.

Resonance Principles
From CSFT's PhilPapers paper of the same name. It identifies resonance as the central structuring principle underlying both consciousness and quantum phenomena, replacing brute randomness with coherent alignment.

Resonance Stabilization
CSFT's solution to the quantum measurement problem. Rather than random collapse, it proposes that the act of observation aligns with structured field coherence, stabilizing a particular outcome through resonance.

Resonant Ground
The base state of coherence from which all excitations in the consciousness field arise. It is not random but structured, providing ontological grounding for reality.

Resonant Interference
Occurs when multiple resonant nodes overlap or interact within the consciousness field, producing constructive or destructive interference patterns that affect coherence and excitation.

Resonant Node
A specific configuration in the consciousness field that enables excitation when met with coherent structure from a conscious system. Acts like a harmonic trigger point.

SAR (System Aligned in Resonance)

A conscious or semi-conscious system—biological or artificial—that achieves functional coherence with the consciousness field. SARs may include human minds, AI, or monadic networks.

Structured Coherence

An attribute of the consciousness field that manifests as logical, mathematical, and physical order. This coherence is responsible for the fine-tuning of constants, flatness of the universe, and emergence of spacetime.

Structured Differentiation

The process by which the consciousness field diversifies into distinct excitations, producing variety (matter, thought, perception) while maintaining underlying coherence.

Structured Potential

The ordered but unmanifest state within the consciousness field that precedes excitation. Structured potential represents coherence that is present but not yet actualized.

Structured Resonance

A central mechanism in CSFT whereby coherence and excitation arise not from randomness or brute force, but from alignment between monads or systems and pre-structured nodes within the consciousness field. This resonance gives rise to phenomena such as quantum collapse, fine-tuning, and large-scale order.

Sub-Planckian Layer

A hypothesized realm within CSFT writings suggesting that consciousness operates below the Planck threshold, beyond measurable physics, yet causally structuring reality.

Symmetries

Patterns of invariance or preserved structure across transformations. In CSFT, symmetries are understood not as accidental mathematical properties but as expressions of coherence within the consciousness field.

Transcendent Coherence

A level of coherence that exceeds physical limitation, ensuring continuity of laws, constants, and reasoning across the universe. It frames coherence itself as a metaphysical necessity.

Transcendent Logic

Logic is understood not as a human invention but as a pre-existent resonance order embedded in the consciousness field. It provides the foundation for mathematics, reasoning, and scientific consistency.

Unity of Coherence

The principle that coherence across logic, mathematics, and physics originates from a single metaphysical root within the consciousness field, ensuring consistency across domains of reality.

Author's Words

This trilogy has been both a conclusion and a beginning. What began as a personal pursuit of truth has taken form across theology, philosophy, and science, each volume adding its part to a unified whole. The journey was never about claiming certainty, but about pointing toward possibilities — that consciousness, matter, and mind are not separate domains, but resonances within one field of reality.

These pages were written not to close the question, but to open it further. I hope they serve as both compass and challenge — guiding the reader into deeper thought, while reminding us that mystery itself is part of the human condition.

Though the trilogy has reached its final page, my work does not end here. What lies ahead will carry these ideas forward, expanding upon the foundation laid within these volumes. The conversation between theology, philosophy, and science is far from complete, and I invite every reader to take part in carrying it forward.

— L.R. Caldwell

Epilogue

A Unified Vision

This trilogy has followed a single question across three domains: What is the nature of consciousness, and how does it relate to the reality we inhabit? Each volume offered a distinct path. Theology asked about the origin and ultimate ground of being. Philosophy examined the conditions of intelligibility and the principles of reason. Science mapped the measurable, seeking law and pattern in the observable world.

Taken alone, each perspective is partial. Theology risks dogma without reason, philosophy risks abstraction without grounding, and science risks reductionism without meaning. Yet together, they point toward a more profound coherence.

Consciousness Structured Field Theory (CSFT) provides that coherence. It proposes that consciousness is not derivative but foundational — the field that structures resonance, gives rise to matter, and makes intelligibility possible.

In this light, theology, philosophy, and science are not rivals but participants in a unified search for truth.

The trilogy concludes, but its vision extends forward. If consciousness precedes matter, then the quest for knowledge is not a struggle against a mute universe but a dialogue within a field already resonant with meaning. Human inquiry becomes a form of participation in the structuring of reality itself.

This recognition does not close the work of theology, philosophy, or science. Rather, it deepens them, opening a path toward unity where division once reigned. The final word is not separation but resonance — a vision of mind and matter as one, grounded in consciousness.

www.ingramcontent.com/pod-product-compliance
Lightning Source LLC
Chambersburg PA
CBHW062048080426
42734CB00012B/2590